# Dedication

A friend whose wisdom surpassed extra length such that extensive riddles

To my mother Ms Dorah Sello and father Mr. NR Sello

Thank you

Mr. T.V Sello

## Preface

As simple as an order of numbering can be by all definitions of theories established by aspects of physical life "everything in life is designed in order such that the probability of configuration is possible". Life is driven mainly by order of behaviors, hence we say "science" or "mathematics".

We study numbers to measure or predict the behavior of future mediums (e.g. whether) to an extent that we use math's to make our life easier. It is then when I introduce this fundamental book to make functions more philosophical real relative to physical life

Hope you find this theorems handful in your career.

Kind regards
Thapelo Vincent Sello

**Researched by**

**company**® **Mathematical**

*Research institute of physicists*

**495 Mashimong section**
**Tembisa**
**1632**
**South Africa**

E-mail: engineering.mathematics@Live.co.za
Facsimile: 086 560 7721

All right reserved. No part of this book may be re-published, reproduced and transmitted in any form or by any means, electronic, mechanical, photocopying, recording or other forms otherwise without prior written permission of the mathematical company.

Copyright © 2011

First impression 2011

## About the book theoretical

Around the world there are many aspects of mathematical methods used to conquer problems associated with numbers which differ along with its theories to this book. The contents of this book are mainly first to book fundamentally and theoretical. The mission of this book is not the same as other books because it does not use $\pi$ as a constant at any form.

Hence the theorems of the book are mainly first to be published within the history of mathematics. The life of mathematics is not complex nor difficult but as easy as possible.

In conclusion this book is theoretical and not practical. And it covers CALCULUS only.

Thank you for your support!

| Contents | Page |
|---|---|

- ♣ Algebra ............................................................................................................ 1
    - History
    - Division of a polynomial

- ♣ Functions........................................................................................................ 8
    - One to one function
    - Many to one function

- ♣ Trigonometry................................................................................................ 18
    - Double angles definitions

- ♣ Calculus ....................................................................................................... 23
    - Gradient
    - Tangent definition
    - Algebraic calculus
    - Trigonometric calculus

- ♣ Inverse functions within calculus
    - Limits

- ♣ Exponential calculus.................................................................................... 52
- ♣ Applications within calculus........................................................................ 56
- ♣ Advancement of analytical geometry and measurement............................ 69
- ♣ Techniques of integration............................................................................. 73

# Notes

# *Algebra*

**In this unit you will learn**

- ✓ The history of algebra
- ✓ The division of a polynomial

However it is important to know where algebra comes from and it purpose as a tool in calculation. The motive and the goal set by this concept of this study is put to light within its history.

## The algebra

The term algebra was derived from the Arabic word called al-jab meaning transposition. Algebra in turn became the branch in mathematics involving cancellations and transpositions. It is still on-going debate over who invented algebra; however several mathematicians below contributed / invented algebra by part of what we study today.

- ✓ **Diaphanous**

  Greek mathematician who invented Diophantine equations.

- ✓ **Al Khwarizmi**

  Arad mathematics responsible for expanding algebra and transmitting Diophantine idea to the western world. He later invented the rules of restoration and reduction.

- ✓ **Leonard Fibonacci**

  Italian mathematician who introduced the Hindu-Arabic numeral (i.e.1, 2, 3,)
  In conjunction with algebra.

- ✓ **Girolamo cordano and Niccole tartaglia**

  They produced cardano's rule which was first used to solve $3^{rd}$ degree equation

- ✓ **François Viete**

  He also developed the algebraic symbolism, using letters to represent unknown variables (today it is known as solve for $x$). He then later advanced algebra). He then later advanced algebraic theory of notation.

- ✓ **John Wallis**

  Invented calculating prodigy

- ✓ **Everest Galois**

  Advanced the grouping theory

- ✓ **Boolean : Gorge Boole and Alfred north Whitehead**

  They invented local algebra which today is used for computers.

## Language within the study

The most important aspect of reading the literature used in this context of analysis of algebra is based in the understanding of quantities. However even primitive people used to speak the language of algebra by reading the quantity of they life stock. Hence today is more advanced so as to credit the name of mathematics tool of language.

## Understanding the basic of the language

- Convert the following scenario into an equation and find the minimum value of the variable and the maximum value of it.

  e.g.

  A kohekohe livestock farmer has cows that which multiply themselves every year. He then realizes that wolfs steal 4 cows from him every year.

**Solution**

If a Cow = x

∴ $x.x$ ⟶ they multiply themselves

Then 4 cows are stole

∴ $x.x - 4$ is a scenario

Therefore $S(x) = x^2 - 4$ ........ an equation

The minimum value is $S'(x) = 2.x^{2-1} - 0 = 0$

$$0 = 2x$$

∴ $x = 0$ ..........substitute it back to the equation

$S'(x) = (0)^2 - 4$

The minimum is -4 and the maximum value is 4 of cows

---

The other one is finding the unknown variable, but first you should understand that

- Multiplication yields products

  e.g. $x.x = x^2$ ⟶ product

  **Multiplication**

- Division also yields products
- Addition and subtraction yield sums in general

---

**Passport to algebra**

<u>In this unit you will learn</u>

- ✓ Polynomials in relation with reminder theorem and synthetic division
- ✓ The use if reminder theorem
- ✓ Factorization with the 3<sup>rd</sup> degree polynomial
- ✓ Euclidean property and fractions

- ✓ Simplification of simultaneous equations

## Polynomials

The historical period known as hindu-arabic from 500 to 1199 AD. The earliest Indian civilization recorded in ruins of the ancient city at Mohenjo Daro in the Indus River, and around 2000B.C. The Aryans established themselves and projected the Sanskrit language and caste system. Hindu astronomer ARYBHATA wrote a mathematical verse that gave rules of solving equations and finding of square to cube roots. It was until BRAHMAGUPTA within Hindu mathematics, whose work was put into book and studied for centuries.

A polynomial is a sum of uncommon terms that are three or more.

## Division of a polynomial

### Fractions

Any number that has a numerator and a denominator.

e.g. $\frac{x}{y}$  (Numerator, denominator)

However given a fraction $\frac{14}{3}$ if you simplify you yield

$4\frac{2}{3}$   Which is a mixed fraction ∴ $4 + \frac{2}{3}$ (reminder)

(Quotient, divisor)

The above means that if $f(x) = 14$ and $g(x) = 3$ are non-zero polynomials where $f(x)$ degree > the degree of $g(x)$, therefore a definition becomes

$$\frac{f(x)}{g(x)} = a(x) + \frac{r(x)}{g(x)}$$

*Function*, quotient, Reminder, divisor

This concept is called **Euclidean property**

Definition

While studying Euclidean property a reminder theorem was established

- ✓ Procedure of reminder theorem about factorization
1. Find the divisor
2. Obey the law of BODMAS.

## Skills show

a) If $f(x) = x^3 + x^2 + x + 1$
   i. Factorize

    ii. **Solve for *x***

b) **If** *f(x) = x³ + x² - 8x – 12*
    i. **Solve for *x***

c) **Given** *g(x) = x³ + 8x² + x + 42*
    i. **Solve for *x***

---

# Functions
# And inverse functions

In this unit you should learn the following

- ✓ One to one functions
- ✓ Many to one functions
- ✓ Inverse functions and it purposes

This study must help you configure valid definable and invalid undefined functions. The purposes of inverse function relative to their existence.

---

### Functions

The other word for function is the graph; however for a function to exist "*y*-axis and *x*-axis" should be present as independent and depended variables. Look at the following diagram called a Cartesian plane

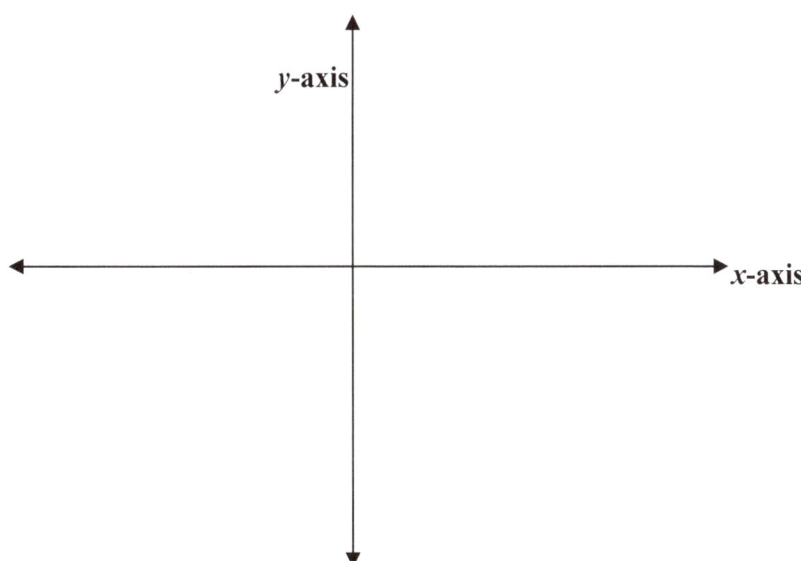

Hence for every function there is a domain(*x*-int) and a range(*y*-int) for which when they combine they make or form a co-ordinate ($x$, $y$). At the later stage co-ordinates construct the presentation of the graph or a function.

### In this unit you will

1. Know the fundamentals of a function.

2. Identify different functions and simplify them with theoretical reasoning

3. Know how to generate and solve graphs of inverse functions.

4. Understand restrictions of a function and its purpose.

5. Plot and interpret functions using critical points.

<u>**Why study?**</u>

To put information in a good form where analysts conclude the information and results of any event we need functions.

## How to know that a graph is regarded as a valid definable funtion

**Remember that a function is constructed by (*x*-int) and (*y*-int) which together produce a co-ordinate. The way of proving if a presentation of line is a function is called vertical line test.**

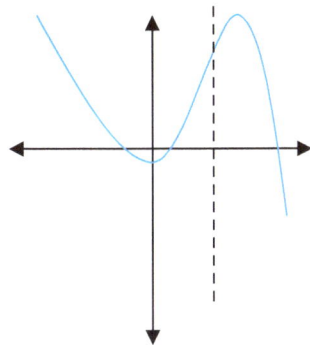

**Fig: A domain touching this graph once**

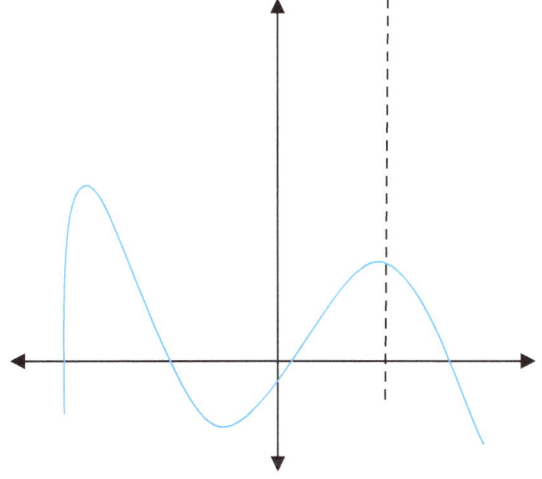

**Fig: domain touching this graph once**

**Fig: a domain touching it twice and more**

However a function a mathematical sensible graph must have a domain cutting through it once, then we can say is a true function. If a domain cuts it many times but at different intervals, the domain is then considered to de infinite i.e.

$$x = \infty \text{ and } y = \infty$$

---

### Range test

Remember that a range is defined by(*y*-int) therefore if the horizontal line once.

E.g.

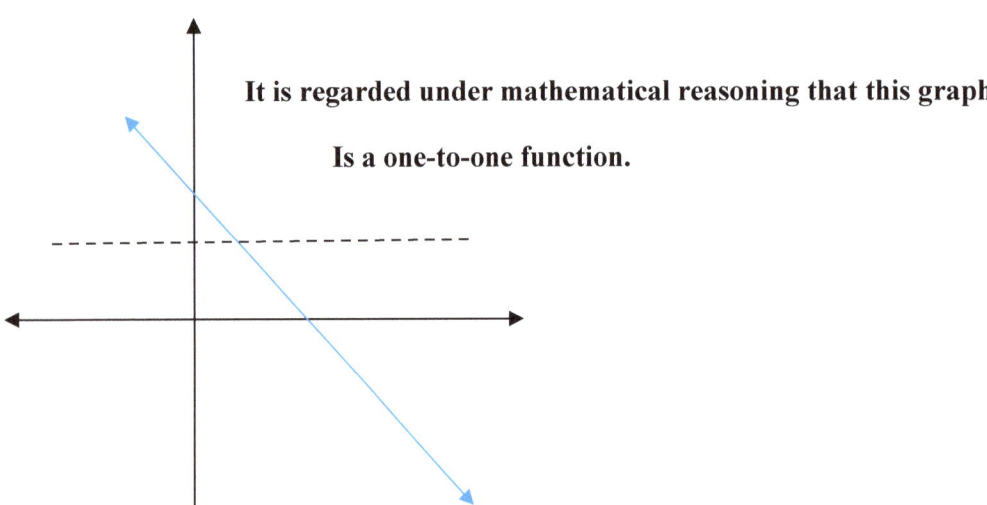

It is regarded under mathematical reasoning that this graph

Is a one-to-one function.

Fig: A graph for one-to-one function.

If a horizontal line cuts a function twice or more, then the function is classified as many to one function. E.g.

The domain is defined infinite because many-to-one functions have many, many, many intervals where it cuts once.

Look at this figure.

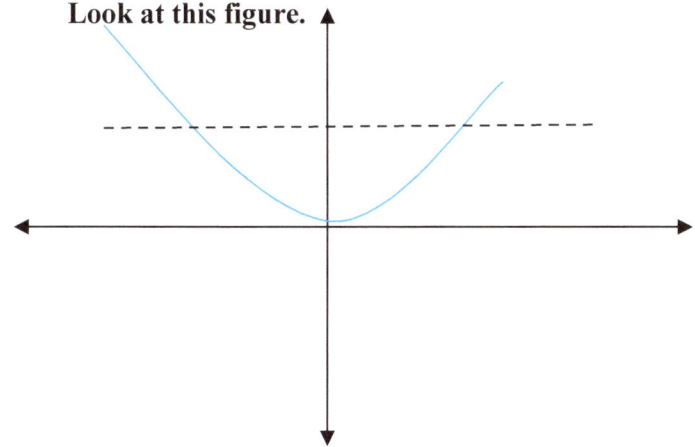

Fig: A many-to one function

### Using calculations

To check if the graph cuts the range at a single point once.

E.g.

$$f(x) = x^2$$

At the point $x = \pm 4$

$$f(4) = (4)^2$$
$$= 16 \quad \ldots\ldots(a)$$
$$f(-4) = (-4)^2$$
$$= 16 \quad \ldots\ldots\ldots(b)$$

The function has the same $y$-int as range, therefore this cuts it once meaning it's a once to one function.

### Skills show

1. (a) If $f(x) = 4$

(i) Sketch the function of $f$.

(ii) Determine if one-to-one function is?

(iii) Is this a function?

(b) Given $f(x) = x^2 + 3x - 10$

(i) Determine the domain and the range of this function.

(ii) Sketch the graph

(iii) What type of a function is this function classified?

2. $P(x) = \dfrac{1}{x+1}$

(i) Are all values of x definable?

(ii)

(iii) What can you conclude?

## Skills show solutions

1. (i)

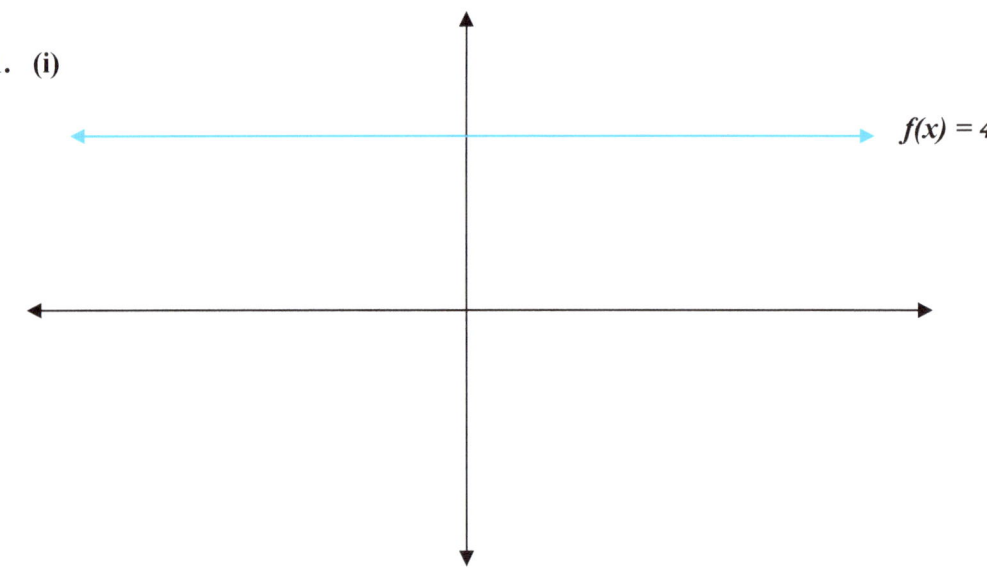

(ii) $f(x) = 4$

$f(-1) = 4$

……..therefore if $f(1) = 4$, the value of $f$ is the same with <u>x-int that has different signs</u> meaning that

**This is one-to-one function**

**With the range of 4**

(b)  (i) $f(x) = x^2 + 3x - 10$

To find x-int let $f(x) = 0$

Therefore $x^2 + 3x - 10 = 0$

$(x + 5)(x - 2) = 0$

$x + 5 = 0$ or $x - 2 = 0$

$x = -5$ or $x = 2$

Then to find y-int, let $x = 0$

$f(x) = x^2 + 3x - 10$

$= (0)^2 + 3(0) - 10$

$= -10$

**Turning point { T.P}**    $x = \dfrac{-b}{2a}$    …….. formulae

Therefore………………

$x = \dfrac{-(3)}{2(1)}$

$x = -\dfrac{3}{2}$

$f(x) = x^2 + 3x - 10$

$f(x) = \left(-\dfrac{3}{2}\right)^2 + 3\left(-\dfrac{3}{2}\right) - 10 = -\dfrac{98}{8}$

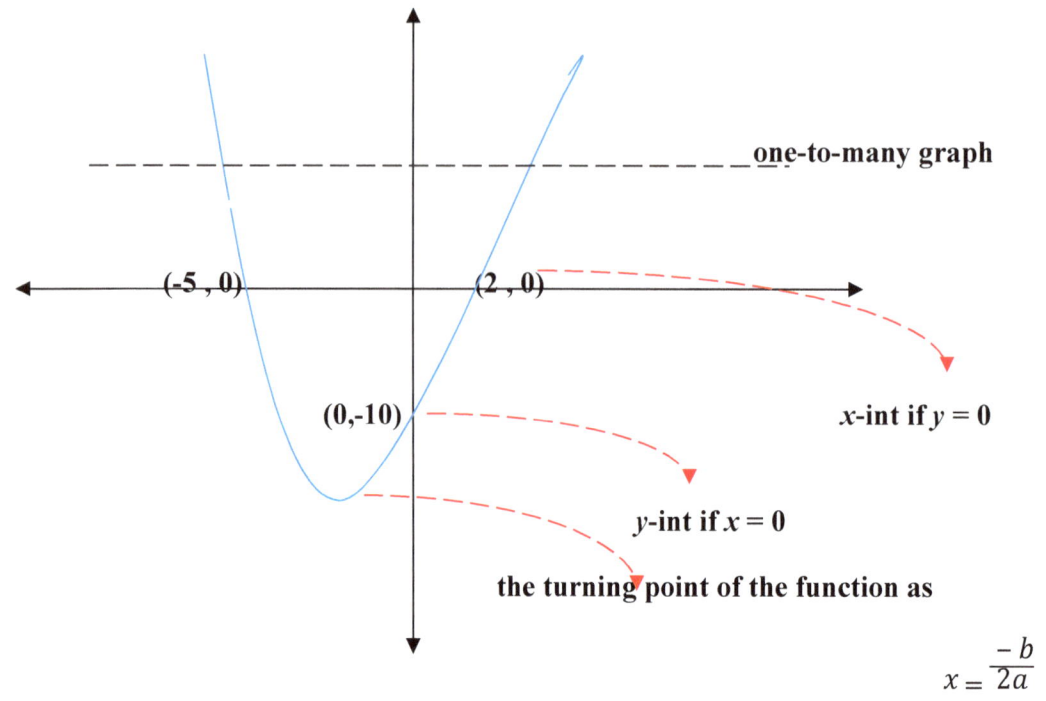

(c) (i) No, because if $x = -1$ therefore $f(x) = \dfrac{1}{(-1)+1} = $ *undefined*.

(ii) Sometimes we have to restrict function so that they become definable.

### In this unit you should

- Understand the definition of a function.
1. Know the domain and its purpose.
2. Also understand one-to-one function and one to many
3. interpretation

### Verify your skills

### Work practice

1. Determine with calculations, which of the following functions are one-to-one functions and one-to-many
   a) $f(x) = \sqrt{x^2 - 1}$
   b) $f(x) = \sqrt{x} + \sqrt[3]{x}$
   c) $g(x) = \dfrac{1}{x + 2}$
   d) $f(x) = x^4$
   e) $f(x) = \sqrt[3]{x^2}$

2. $B(x) = \dfrac{1}{x + 4} + 1$
   a) For which values of $x$ is the function undefined
   b) What can you conclude

**Inverse functions**

At the end of any debate within the word inverse you will find that both parties agree on the word opposite.

However if

$$4 \longrightarrow \text{transforms to} \longrightarrow 8$$

That concludes that its opposite is therefore

$$8 \longrightarrow \text{trnsforms back to} \longrightarrow 4$$

Hence will study functions with their opposite function

Imagine a function of $f(x) = x^2 + 1$ such that when $x = 3$ is substituted, the function yields 10.
   Therefore if you want to reverse 10 to 3, then you will need an opposite function whereby it is defined by $f^{-1}(x)$ as it`s denote[symbol]. Imagine given the same function

### Skills show

$f(x) = x^2 + 1$  ……….. simply by making $x$ the subject of the function
$\therefore x^2 = f(x) - 1$
$x = \sqrt{f(x) - 1}$

$f(x) = \sqrt{x-1}$ ...... interchange the variables
$f^{-1}(x) = \sqrt{x-1}$ ........ subtitute the inverse denote

To prove that this an inverse function
$f(f^{-1}(x)) = (\sqrt{x-1})^2 + 1$
$\quad\quad\quad = x$ .....therefore this is an inverse function

### In trigonometry
It is still the same definition .

### Skills show
e.g. if $f(x) = \sin x$
$$\therefore x = \sin^{-1} f(x)$$
$$f^{-1}(x) = \sin^{-1} x$$
to prove
$f(f^{-1}(x)) = \sin \sin^{-1} x$
$\quad\quad\quad = x$

---

a) Determine inverse /opposite functions of the following and which one doesn`t have an inverse
    (i) $f(x) = 4x + 1$
    (ii) $f(x) = x^2 + 2x + 1$
    (iii) $f(x) = \dfrac{9x+3}{2}$

b) Prove that $g(x)$ is an inverse function of $f(x)$.
    (i) $g(x) = x - 1$      ; $f(x) = x + 1$
    (ii) $g(x)\ 7x - 1$      ; $f(x) = \dfrac{x+1}{7}$
    (iii) $g(x) = \sin^{-1} x$      ; $f(x) = \sin x$
    (iv) $g(x) = \tan^{-1} x$      ; $\tan x = f(x)$

### Skills show solutions
(i) If $f(x) = 4x + 1$
$$x = \dfrac{f(x) - 1}{4}$$

$$f^{-1}(x) = \frac{x-1}{4}$$

(ii) If $f(x) = x^2 + 2x + 1$

$f^{-1}(x)$ does not exist because it a one-to-many function.

(iii) $f(x) = \frac{9x + 3}{2}$

$2f(x) = 9x + 3$

$x = \frac{2f(x) - 3}{9}$

$f^{-1}(x) = \frac{2x - 3}{9}$

b) Definition states that $f(f^{-1}(x)) = x$

(i) $f(g(x)) = (x - 1) + 1$
   $= x$ ……..therefore it is an inverse function.

(ii) $f(g(x)) = \frac{(7x - 1) + 1}{7}$
   $= x$

(iii) $f(g(x)) = \sin \sin^{-1} x$
   $= x$

(iv) $f(g(x)) = \tan \tan^{-1} x$
   $= x$

## Verify your skills

1. Show that $g(x)$ is an inverse of $f(x)$.

   a) $g(x) = 2x + 1$ ; $f(x) = \frac{x-1}{2}$
   b) $g(x) = \sin^{-1}(x + 1)$ ; $f(x) = \sin x + 1$
   c) $g(x) = \sqrt[5]{x - 3}$ ; $f(x) = x^5 + 3$

## Conclusion of functions

On the same set of $f(x)$ it is an inverse $f^{-1}(x)$, $f(x)$ is then reflected by the line $y = x$ to become $f^{-1}(x)$.

Given $g(x)$ as a inverse function of $f(x)$

$$F(g(x)) = x$$

Meaning a domain ( *x-int* ) of $f(x)$ is the same /equal to the range of $g(x)$. And vise-visor.

Only one-to-one functions have an inverse function

___

# Trigonometry

**In this unit you must understand**

- ✓ Understand double angles
- ✓ Basic trigonometry within Pythagoras theorem

## Trigonometry

It is known as the study of properties and proportions of triangles. It was founded in the Greece around the third century B.C {3000B.C}. It was developed because accurate astronomical calculation was later successful as an aid in geography.

- ✓ Map making
- ✓ Surveying
- ✓ Positioning
- ✓ Direction in aviation industry

Hence the spherical trigonometry is used to determine

- ✓ The time of the day
- ✓ Direction of motion

## Some scientist like the following

- ✓ **Menelaus (astronomer)**
  Developed spherical trigonometry
- ✓ **Ptolemy**
  Construction of trig ratio tables
- ✓ **Hindus and Arabs**
  Determined the tangent

All of them created what we study today we call trigonometry and recent mathematicians only advanced the field.

## In this unit you will

- ✓ You will derive identities
- ✓ Also use double angles and its identities
- ✓ Master general solutions
- ✓ Learn how to solve trigonometric equations using compound angle identities

## Deriving compound angles and their identities by geometrical aspect

**Pythagoras theorem states that**

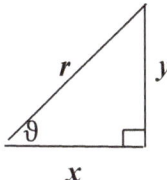   ∴ $x^2 + y^2 = r^2$     …..theorem

therefore

$\sin \vartheta = \dfrac{y}{r}$   ……  ∴ $y = r \sin \vartheta$

$\cos \vartheta = \dfrac{x}{r}$   ……  ∴ $x = r \cos \vartheta$

*using the Pythagoras theorem*

$(r \sin \vartheta)^2 + (r \cos \vartheta)^2 = r^2$

$\dfrac{r^2 \sin^2 \vartheta}{r^2} + \dfrac{r^2 \cos^2 \vartheta}{r^2} = \dfrac{r^2}{r^2}$

**therefore the fundamental identity is born stating that**

$$\sin^2 \vartheta + \cos^2 \vartheta = 1$$

basic identity

later in this chapter you will realize that this identity is the mother of all identities , however despite other complex functions , you will also have to master this identity so as to form your own identity as a mathematician.

Now imagine given the sum of the angles in a right angled triangle!!

## Sum of the angles called double angles

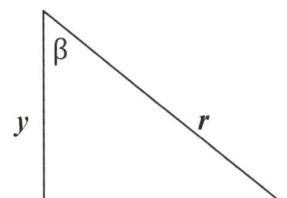

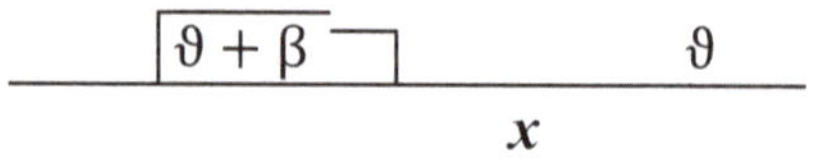

```
|  ϑ + β  |         ϑ        |
             x
```

let's agree that :

$\sin ϑ = \frac{x}{r}$

$\cos ϑ = \frac{y}{r}$         and    $\cos β = \frac{x}{r}$

$\sin β = \frac{y}{r}$

*Looking carefully you will see that:*

Sin ϑ = cos β

Cos ϑ = sin β         ∴ $Sin^2 ϑ + Cos^2 ϑ = 1$

Sin . Sin ϑ + Cos . Cos ϑ = 1

therefore substitute    Cos β . Sin ϑ + Sin . Cos ϑ = 1

Sin 90 = 1

Sin ( + β) = 1

Therefore a definition for double angles is

| Cos . Sin ϑ + Sin . Cos ϑ = sin ( ϑ + β) |

definition 1.1

| Cos . Sin ϑ - Sin . Cos ϑ = Sin ( ϑ - β) |

Definition 1.2

However this is a sine double angle definition


## Concluding the definitions

- ✓ $\sin^2 ϑ + \cos^2 ϑ = 1$ is a basic identity in trigonometry
- ✓ cos . sin ϑ - sin β . cos ϑ = sin ( ϑ - β) is the sine double angle identity with -
- ✓ cos . sin ϑ + sin β . cos ϑ = sin ( ϑ + β) is the sine double angle identity with +

the above proven using Pythagoras theorem

- ✓ cos (β + ϑ) = cos ϑ . cos β - sin ϑ . sin β
- ✓ cos (β - ϑ) = cos ϑ . cos β + sin ϑ . sin β

## tangent

**Sum of the angles**

$$\tan(\vartheta + \beta) = \frac{\sin(\vartheta + \beta)}{\cos(\vartheta + \beta)} = \frac{\sin\vartheta \cos\beta + \sin\beta \cos\vartheta}{\cos\vartheta \cos\beta - \sin\vartheta \sin\beta} \times \frac{\frac{1}{\sin\vartheta}}{\frac{1}{\sin\vartheta}}$$

$$= \frac{\frac{1}{\sin\vartheta}(\sin\vartheta \cos\beta + \sin\beta \cos\vartheta)}{\frac{1}{\sin\vartheta}(\cos\vartheta \cos\beta - \sin\vartheta \sin\beta)}$$

$$= \frac{(\cos\beta + \frac{\sin\beta}{\tan\vartheta})}{(\frac{\cos\beta}{\tan\vartheta} - \sin\beta)} \times \frac{\frac{1}{\cos\beta}}{\frac{1}{\cos\beta}}$$

$$= \frac{\frac{1}{\cos\beta}(\cos\beta + \frac{\sin\beta}{\tan\vartheta})}{\frac{1}{\cos\beta}(\frac{\cos\beta}{\tan\vartheta} - \sin\beta)}$$

$$= \frac{1 + \frac{\tan\beta}{\tan\vartheta}}{\frac{1}{\tan\vartheta} - \tan\beta}$$

$$= \frac{\tan\vartheta + \tan\beta}{\tan\vartheta} \times \frac{\tan\vartheta}{1 - \tan\vartheta \tan\beta}$$

$$= \frac{\tan\vartheta + \tan\beta}{1 - \tan\vartheta \tan\beta}$$

**This are double angles that you can study as far as this study is concerned. Study these identities as the passport to trigonometry.**

Using the above definitions you can solve any problem that is presented to you irrespective of the complexity it holds closing basic trigonometry at this note, the skills of problem solving depends on approach of self intelligence. This is because of the way trigonometry present itself. On the topic of identities, it depends on what you want you want to make.

# *Calculus*

### Partial differentiation

### In this unit you will understand

1. The standard first principle of differentiation and limits
2. Limits of trigonometrical equations
3. Concept of a derivative
4. Functions
5. Inverse functions
6. Rates of change
7. Identities

### Gradient

It is the steepness of the slope.

e.g.        **Increasing**          **Stationary**          decreasing

$\tan \vartheta$ = gradient = m          ………formulae (m is a gradient)

Imagine given the curve to find a gradient.

$f(x) = x^2$

On the curve, a gradient at every point changes.
assume that $g(x)$ touches the curve once
And at w it touches the curve.

$g(x) = mx + c$

### Ask yourself the following ……

1. At point w does the graph increase or decrease?
2. What is the gradient of *g*?

## Limits

Previously on functions we dealt with restrictions of a function.

e.g. $f(x) = \dfrac{1}{x+1}$     restriction is $x < -1 \text{ and } x > -1$

Instead of restrictions within calculus we say limits.

## Skills show

If $f(x) = \dfrac{x^2 - 1}{x + 1}$

    i.       When $x = -1$ does $f$ become definable?
    ii.      How can you make $f$ definable?
    iii.     Hence show the value of $f$ proved that $x = -1$
    iv.     What is the limit for $f$
    v.      Therefore $f(x) = \ldots\ldots\ldots$ as x approaches $\ldots\ldots\ldots$
    vi.     Express $f(x)$ in a concluded mathematical form

Given $f(x) = 7x + 1$

    i.       Determine the gradient
    ii.      Does it have limits? Explain

## Skills show solutions

    i.       If $f(x) = \dfrac{x^2 - 1}{x + 1}$

Therefore $f(-1) = \dfrac{(-1)^2 - 1}{-1 + 1} = \dfrac{0}{0} = $ undefined

ii. Simplifying $f$

$$f(x) = \frac{x^2 - 1}{x + 1} = \frac{(x+1)(x-1)}{x+1} = x - 1$$

iii. $f(x) = x - 1$
therefore $f(-1) = (-1) - 1$
$= -2$

iv. the limit is -1

v. $f(x) = -2$ as $x \to -1$

vi. $f(x) = \lim_{x \to -1} \frac{x^2 - 1}{x + 1}$

vii. Given $f(x) = 7x + 1$

i. $y = mx + c$ ..........*is the formulae of which* $y = 7x + 1$
*therefore m = 7*

ii. no. because its linear

## Conclusion

It is then seen by solving the above challenges that:

1. Tangents touch a curve at specific points.
2. Limits can be definable
3. Restrictions are important so as to make functions definable

### At the end of this unit you should :

1. Understand derivativeness of limits
2. Tangents to a curve relative to gradients

**Verify your skills**

**Limits**

a) $\lim\limits_{h \to 0} \dfrac{x^2 + 4x\, 4}{x + 2}$

b) $\lim\limits_{h \to 0} \dfrac{x^2 - 4x + 4}{-x + 4}$

c) $\lim\limits_{h \to 0} \dfrac{x^3 + x^2 + x + 1}{x + 1}$

d) $\lim\limits_{h \to 0} \dfrac{(x+h)^{-1} - x^{-1}}{h}$

e) $\lim\limits_{h \to 0} \dfrac{2^{x-1} - 0.5}{0.5}$

f) $\lim\limits_{h \to 0} \dfrac{f(3+h) - f(x)}{h}$  if $f(x) = 6x$

g) $\lim\limits_{h \to 0} \dfrac{f(x+h) - f(x)}{h}$  if $f(x) = c$

h) $\lim\limits_{h \to 0} \dfrac{f(x+h) - f(x)}{h}$  if $f(x) = \dfrac{1}{\sqrt{x}}$  {hint:rationalize}

## Conclusion

- Limits do not exist at the linear definition
- Limits are expressed as $\lim_{h \to 0} \dfrac{f(x+h) - f(x)}{h}$

## Differentiation

Within the chapter of calculus, it is known by mathematicians that so far

> $f(x)$ is differentiable at $x$ if at point $(x, f(x))$ a
> 
> $\lim_{h \to 0} \dfrac{f(x+h) - f(x)}{h}$  Exist.

Definition

The above principle is therefore called the <u>first principle of differentiation</u>

## Skills show

a) Use the above principle to determine $f'(x)$
1) $4x^2$
2) $3x - 1$
3) $2x^2 + x$
4) $3$
5) $x + 2$

b) Hence conclude the knowledge.

## Skills show solutions

**Implying**

a) $$f'(x) = \lim_{h \to 0} \frac{f(x+h) - f(x)}{h}$$

$$= \lim_{h \to 0} \frac{4(x+h)^2 - 4x^2}{h}$$

$$= \lim_{h \to 0} \frac{4x^2 + 8xh + 4h^2 - 4x^2}{h}$$

$$= \lim_{h \to 0} \frac{h(8x + 4h)}{h}$$

$$= \lim_{h \to 0} 8x + 4h \quad \ldots\ldots\text{subtitude } h$$

$$= 8x + 4(0) = 8x$$

**Try the rest.**

## Verify your skills

**First principle of differentiation**
**Find the limits of the following functions**

a) $4x^3$
b) $\dfrac{1}{\sqrt{x+1}}$     *{hint : rationalize}*
c) $x^3 - 1$
d) $x^4$
e) $\dfrac{x^2 - 1}{x}$

## Derivatives

It is the gradient at a specific point. It further concludes to state

That

---
$f(x)$ at the point $(x, f(x))$, has the gradient of $\displaystyle\lim_{h \to 0} \frac{f(x+h) - f(x)}{h}$ (if it exist)

---

**Definition**

## Skills show

**Find the gradient of a curve defined by**

a) $f(x) = x^2$ at $x = 3$
b) $f(x) = x^4 + x^3$ at $x = -2$
c) $f(x) = 3x^2$ at $x = -2$
d) $f(x) = 2x^2 - x + 1$ at $x = 0$

## Skills show solution

a) $f'(x) = \displaystyle\lim_{h \to 0} \frac{f(x+h) - f(x)}{h}$     $\therefore f'(x) = 2x$

$= \displaystyle\lim_{h \to 0} \frac{(x+h)^2 - x^2}{h}$     $f'(3) = 2(3)$

$= \displaystyle\lim_{h \to 0} \frac{x^2 + 2xh + h^2 - x^2}{h}$

$= 6$

$= \displaystyle\lim_{h \to 0} \frac{h(2x + h)}{h}$

$= \displaystyle\lim_{h \to 0} 2x + h$     .....subtitude $h$ by 0

$= 2x + (0) = 2x$

$f'(x)$ *is the gradient at the specific point and hence the definition is used to solve the above problem.*

b) $$f'(x) = \lim_{h \to 0} \frac{(x+h)^2(x+h)^2 + (x+h)^2(x+h) - (x^4 + x^3)}{h}$$

$$= \lim_{h \to 0} \frac{(x^2 + 2xh + h^2)(x^2 + 2xh + h^2) + (x+h)(x^2 + 2xh + h^2) - x^4 - x^3}{h} \quad simplify$$

$$= \lim_{h \to 0} \frac{x^4 + 2x^3h + x^2h^2 + 2x^3h + 4x^2h^2 + 2xh^3 + x^2h^2 + 2xh^3 + h^4 + x^3 + 3x^2h + 3xh^2 + h^3 - x^4 - x^3}{h}$$

$$= \lim_{h \to 0} \frac{h(4x^3 + 6x^2h + 4xh^2 + h^3 + 3x^2 + 3xh + h^2)}{h}$$

$$= \lim_{h \to 0} 4x^3 + 6x^2h + 4xh^2 + h^3 + 3x^2 + 3xh + h^2 \quad \text{.......substitute } h \text{ by } 0..$$

$$= 4x^3 + 6x^2(0) + 4x(0) + (0)^3 + 3x^2 + 3x(0) + (0)^2$$
$$= 4x^3 + 3x^2$$

$\therefore f'(-2) = 4(-2) + 3(-2)$
    $= -20$

The ultimate trick of the above expression is based on how to
- ✓ Factorize
- ✓ Indentifying like terms
- ✓ Inspection within factorization

Try to do the rest

## Conclusion

From what you have learned you should have the ability to

- ✓ **Know how to differentiate**
- ✓ **Understand limits**
- ✓ **Solve problems involving tangents to a curved function**
- ✓ **Use derivatives**
- ✓ **Analyze gradients**
- ✓ **Know the concept of a curve**

## Trigonometric calculus

In this unit we will further the knowledge of limits within

- ✓ Sine function
- ✓ Cosine
- ✓ Tangents

However the learner should first understand the basics of

- ✓ Basics of trigonometry
- ✓ Double angles

## Trigonometric limits

Remember that limits are referred as restrictions before differentiation

Given $f(x) = \sin ax$

a) $$\therefore f'(x) = \lim_{h \to 0} \frac{f(x+h) - f(x)}{h}$$

$$= \lim_{h \to 0} \frac{\sin a(x+h) - \sin ax}{h}$$

If $x = 0$ therefore

$$f'(0) = \lim_{h \to 0} \frac{\sin a(0+h) - \sin a(0)}{h}$$

$$= \lim_{h \to 0} \frac{\sin ah}{h} = a$$

Therefore the sine limit is defined as

$$\lim_{h \to 0} \frac{\sin ah}{h} = a \quad \text{is the limit of a sine function}$$

definition

Given $f(x) = \cos ax$

$$f'(x) = \lim_{h \to 0} \frac{f(x+h) - f(x)}{h}$$

$$= \lim_{h \to 0} \frac{\cos a(x+h) - \cos ax}{h}$$

$$\therefore f'(0) = \lim_{h \to 0} \frac{\cos a(0+h) - \cos a(0)}{h}$$

$$= \lim_{h \to 0} \frac{\cos ah - 1}{h} = 0$$

Therefore the cosine limit is defined as

$$\lim_{h \to 0} \frac{\cos ah - 1}{h} = 0$$

As the function is being differentiated

## Tangent

Given $f(x) = \tan ax$

$$f'(x) = \lim_{h \to 0} \frac{f(x+h) - f(x)}{h}$$

$$= \lim_{h \to 0} \frac{\tan a(x+h) - \tan ax}{h}$$

$$\therefore f'(x) = \lim_{h \to 0} \frac{\tan a(0+h) - \tan a(0)}{h}$$

$$= \lim_{h \to 0} \frac{\tan ah}{h} = a$$

However the tangent's limit is defined as

$$\lim_{h \to 0} \frac{\tan ah}{h} = a \text{ as it becomes differentiated}$$

**Definition**

## Skills show

**Determine the limits of the following if they exist and conclude what you realize.**

a) $f(x) = \sin x$
b) $f(x) = \cos x$
c) $f(x) = \tan x$

## skills show solution

a) $f'(x) = \lim_{h \to 0} \dfrac{f(x+h) - f(x)}{h}$

$= \lim_{h \to 0} \dfrac{\sin(x+h) - \sin x}{h}$ ......use double angle defination

$= \lim_{h \to 0} \dfrac{\sin x \cdot \cos h + \sin h \cdot \cos x - \sin x}{h}$ ....common factor

$= \lim_{h \to 0} \dfrac{\sin x (\cos x - 1) + \sin h \cdot \cos x}{h}$ ........ $\dfrac{a+1}{a} = \dfrac{a}{a} + \dfrac{1}{a}$ ......remember

$= \lim_{h \to 0} \dfrac{\sin x (\cos x - 1)}{h} + \dfrac{\sin h \cdot \cos x}{h}$ ...............subtitude the limits

$= \sin x \cdot \lim_{h \to 0} \dfrac{\cos h - 1}{h} + \cos x \lim_{h \to 0} \dfrac{\sin h}{h}$

$= \sin x \cdot 0 + \cos x \cdot 1$

$= \cos x$

b) $f'(x) = \lim_{h \to 0} \dfrac{f(x+h) - f(x)}{h}$

$= \lim_{h \to 0} \dfrac{\cos(x+h) - \cos x}{h}$

$= \lim_{h \to 0} \dfrac{\cos x \cdot \cos h - \sin h \cdot \sin x - \cos x}{h}$

$= \lim_{h \to 0} \dfrac{\cos x (\cos h - 1) - \sin h \cdot \sin x}{h}$

$= \cos x \cdot \lim_{h \to 0} \dfrac{\cos h - 1}{h} - \sin x \cdot \lim_{h \to 0} \dfrac{\sin h}{h}$

$= \cos x \cdot 0 - \sin x \cdot 1$

$= -\sin x$

a) $f'(x) = \lim_{h \to 0} \dfrac{f(x+h) - f(x)}{h}$

$= \lim_{h \to 0} \dfrac{\tan(x+h) - \tan x}{h}$

$= \lim_{h \to 0} \dfrac{\dfrac{\tan x + \tan h}{1 - \tan x . \tan x} - \tan x}{h}$

$= \lim_{h \to 0} \dfrac{1}{h}\left(\dfrac{\tan x + \tan h}{1 - \tan x . \tan h} - \dfrac{\tan x}{1}\right)$

$= \lim_{h \to 0} \dfrac{1}{h}\left(\dfrac{\tan x + \tan h - \tan x(1 - \tan x . \tan h)}{1 - \tan x . \tan h}\right)$

$= \lim_{h \to 0} \dfrac{1}{h}\left(\dfrac{\tan x + \tan h - \tan x + \tan^2 x . \tan h}{1 - \tan x . \tan h}\right)$

$= \lim_{h \to 0} \dfrac{1}{h}\left(\dfrac{\tan h (1 + \tan^2 x)}{1 - \tan x . \tan h}\right)$

$= \lim_{h \to 0} \dfrac{\tan h}{h}\left(\dfrac{1 + \tan^2 x}{1 - \tan x . \tan h}\right)$

$= \lim_{h \to 0} \dfrac{1 + \tan^2 x}{1 - \tan x . \tan h}$

$= \dfrac{1 + \tan^2 x}{1 - \tan x . \tan(0)}$

$= \dfrac{1}{\cos^2 x}$

## Conclusion

**This unit of study concludes that**

- ✓ **Sine function has** $\lim_{h \to 0} \dfrac{\sin ah}{h} = a$ as a limit
- ✓ **Cosine function also has** $\lim_{h \to 0} \dfrac{\cos ah - 1}{h} = 0$ as a limit
- ✓ **Tangent function has a limit of** $\lim_{h \to 0} \dfrac{\tan ah}{h} = a$

✓ We then use this limits to differentiate from the first principle.

## Verify your skills

## Trig. Limits

a)      Differentiate from the first principle
- i.    $f(x) = \sin^2 x$
- ii.    $f(x) = \dfrac{\sin x^2}{2}$
- iii.    $f(x) = \cos x + \sin x$
- iv.    $f(x) = \dfrac{1}{\sin x}$
- v.    $f(x) = \dfrac{6}{\cos 3x^2}$

b) Hence sketch the graph each on the same of axis with their derivatives

c) Simplify the following
- i.    $\lim\limits_{h \to 0} \dfrac{\sin 4(x+h)^2 - \sin 4x^2}{h}$
- ii.    $\lim\limits_{h \to 0} \dfrac{3\cos 2(x+h) - 3\cos 2x}{h}$
- iii.    $\lim\limits_{h \to 0} \dfrac{7\cos(x+h) - \cos(x+h)^2 - (7\cos x - \cos x^2)}{h}$

d) Hence sketch the functions in a) along with the derivatives.

## Limits and derivatives as a tool to apply

This tool is able to do the following

- ✓ Explain events expressed in a form of a function to whether they did good or bad.
- ✓ Determine the minimum value of a function as it may be necessary. e.g. when engineers calculate how much quantity of cement is needed to build the certain area of space, they eventually use this tool to figure out the problem.

- ✓ To calculate the future i.e. the risk, possibilities etc. e.g. accountants use this tool measure the maximum profit and the minimum profit over a certain period of time.
- ✓ Geographers use this tool to predict future earthquakes. E.g. the seismograph records that the earths

## Short-cut to Differentiation

As far as this study of calculus is concern, as functions become bigger and complex we need to simply them in a simpler way which is solving the problems using short-cuts as a weapon. However we also need to understand that functions have the ways of common characterists, we in turn we have to study so that we master they behavior as we differentiate. In the unit we will also study

- ✓ Algebraic calculus
- ✓ Trigonometric calculus

### Algebraic calculus

If $f(x) = x^n$

$$\therefore f'(x) = n \cdot x^{n-1} \quad \text{....short-cut}$$

And provided that the expression is complex like the following

$f(x) = (x^n + a)^k$

$$\therefore f'(x) = k \cdot (n \cdot x^{n-1} + 0) \cdot (x+a)^{k-1}$$
$$= k \cdot n \cdot x^{n-1} (x+a)^{k-1} \quad \text{.....short-cut}$$

Short-cut definition

### Skills show

Find the derivative of the following using the above definition

i. $f(x) = x^2 + x + 1$

ii. $f(x) = \dfrac{1}{\sqrt{x}}$

iii. $f(x) = ax^2 + bx + c$
iv. $f(x) = 3(x-2)^{16}$
v. $f(x) = (x-2)^2$
vi. $f(x) = 9(x^2 + 4x^3)$

## Skills show solution

i. $f'(x) = 2x^{2-1} + 1.x^{1-1} + 0$
 $= 2x + 1$

ii. *Simplify first*

If $f(x) = \sqrt{x} = x^{\frac{1}{2}}$, $\dfrac{1}{x^{\frac{1}{2}}} = x^{-\frac{1}{2}}$

$\therefore f'(x) = -\dfrac{1}{2}.x^{-\frac{1}{2}-1}$

$= -\dfrac{1}{2x^{\frac{3}{2}}}$

$= -\dfrac{1}{2\sqrt[3]{x^2}}$

iii. $f'(x) = 2a.x^{2-1} + b.1.x^{1-1} + 0$
$= 2ax + b$

iv. $f'(x) = 3.1.x^{1-1}16(x-2)^{16-1}$
$= 48(x-2)^{15}$

v. $f'(x) = 2.1.x^{1-1}(x-1)^{2-1}$
$= 2(x-1)$
$= 2x - 2$

vi. If $f(x) = 9(x^2 + 4x^3)^2$
$\therefore f'(x) = 9(2x^{2-1} + 4.3x^{3-1}).2(x^2 + 4x^3)^{2-1}$
$= 18(2x + 12x^2)(x^2 + 4x^3)$

## However the learner should watch out for

- ✓ uncommon factors so as to make sure that they do not mix
- ✓ formulae so that they are always brain known
- ✓ follow the rules of differentiation
- ✓ respect the BODMAS rule

## Verify your skills

a) find the derivatives of the following function
   i. $f(x) = x^{a+b}$
   ii. $g(x) = 6(x^2 + a)^n$
   iii. $a(x) = \dfrac{x}{x^3 + x^2}$
   iv. $h(x) = \dfrac{1}{x+1}$
   v. $D(x) = ax^3 + bx^2 + cx + d$
   vi. $e(x) = 2x + 1$
   vii. $b(x) = x^2 + a^2 + b^2$
   viii. $p(x) = x^{10} - x^9$

## Conclusion

✓ $f'(x) = \lim_{h \to 0} \dfrac{f(x+h) - f(x)}{h} = n \cdot x^{n-1}$

If $f(x) = x^n$

## Short-cut to differentiation trigonometric calculus

Using the first principle of differentiation the following formulae results as short-cuts to differentiate.

### Short-cut to differentiation

- ✓         $f(x) = d\sin^k ax^b$

$$\therefore f'(x) = abdx^{b-1}\cos ax^b \cdot k(f(x))^{1-\frac{1}{k}}$$

- ✓         $f(x) = d\cos^k ax^b$

$$\therefore f'(x) = -abdx^{b-1}\sin ax^b \cdot k(f(x))^{1-\frac{1}{k}}$$

- ✓         $f(x) = d\tan^k ax^b$

$$\therefore f'(x) = \frac{abdx^{b-1}k(f(x))^{1-\frac{1}{k}}}{\cos^2 ax^b}$$

These are the ultimate short-cut that are easy to master as a junior mathematician as far as these study is concerned.

However you are likely to find them very easy!!!!

## Skills show

a)     find the derivatives of the following functions using the definition above

     i.     $f(x) = 3\sin 2x + 1$
     ii.     $f(x) = 7\tan^2 4x^3$
     iii.     $f(x) = 2\cos^{-0.025} 3x^2$

## Skills show solutions

a)     finding the derivatives

     i.     $f(x) = 2\sin 2x + 1$

$$\therefore f'(x) = abdx^{b-1}\cos ax^b \cdot k(f(x))^{1-\frac{1}{k}}$$

$$= 2 \cdot 1 \cdot 3 \cdot x^{1-1}\cos 2x \cdot 1 \cdot (\sin 2x)^{1-\frac{1}{1}}$$
$$= 6\cos 2x$$

**Simple as 1,2,3**

     ii.     If $f(x) = 7\tan^2 4x^3$

$$\therefore f'(x) = \frac{abdx^{b-1}k(f(x))^{1-\frac{1}{k}}}{\cos^2 ax^b}$$

$$= \frac{4 \cdot 3 \cdot 7 \cdot x^{3-1} 2(\tan^2 4x^3)^{1-\frac{1}{2}}}{\cos^2 4x^3}$$

$$= \frac{168x^2 \tan 4x^3}{\cos^2 4x^3}$$

iii. If $f(x) = 2\cos^{-\frac{1}{40}} 3x^2$

$$\therefore f'(x) = -abdx^{b-1}\sin ax^b \cdot k(f(x))^{1-\frac{1}{k}}$$

$$= -3 \cdot 2 \cdot 2 \cdot x^{2-1}\sin 3x^2 \cdot -\frac{1}{40}(\cos^{-\frac{1}{40}})^{1+\frac{1}{1/40}}$$

$$= \frac{-12x\sin 3x^2 \cdot \cos^{-\frac{41}{40}} 3x^2}{40}$$

## Consider this

- ✓ You must know all the variables before substituting
- ✓ Understand basic algebra to solve the problem after substitution

**Verify your skills (differentiation)**

## Exercise your brain

a) Determine the derivative of

i. $f(x) = \frac{1}{\sin x}$

ii. $f(x) = \frac{2}{\cos x}$

iii. $f(x) = \frac{1}{\tan x}$

iv. $f(x) = \frac{1}{\sin x}$

v. $f(x) = 3\sin^2 x$

## At the end of this unit you should

- ✓ know trigonometrical limits

- ✓ understand limits and their definition
- ✓ master the first principle of differentiation ( trigonometrically and generally)
- ✓ know how to use the derivative
- ✓ understand short-cuts to differentiation (esp. trigonometrically)
- ✓ Explore calculus by its definitions.

By knowing the study of limits and partial differentiation you are now a mathematician with a tool of calculus. Some people will underestimate this but it is very important especially within the following fields

- 90% of engineering field
- 60% of accountancy
- 40% in statistics
- 40% in medicine and mechanical medicine field
- 100% in astronomy

However the circumstance of this tool, it is very important that you understand this because unlike other professionals, this tool is also the basic need in technology.

## Apply your skills

### Assignment

$f(x) = x^3 + x^2 - 3$ is a function of a distance from Johannesburg to Pretoria and it also given that $f(t) = t^2 - \sqrt[3]{\frac{1}{t}}$ is a time taken to get at Pretoria from Johannesburg.

1. Determine the minimum velocity they would require
2. Give the maximum velocity they could probably travel with.
3. What your advice to the travelers about their velocity from JHB to PTA.
4. Write a report on possible velocity and state whether it can happen or not . present this in a scientific manner.

Assessment on the whole study so far of calculus with investigations

## Mastering your skills

### Differentiation and limits

a) **Differentiate the following functions**
   i. $g(x) = 4\sin^{29}(\sqrt[3]{x+1})$
   ii. $f(x) = \cos ax + \sin a^2 x$
   iii. $C(x) = x^6 + \sin 2\sqrt{x}$
   iv. $J(x) = \tan x + \dfrac{9}{\cos x}$
   v. $A(x) = 32\tan 3x^{10} + \dfrac{1}{x}$
   vi. $f(x) = 7x - 3\cos x - 5\sin x^3 + 2\tan x$

b) The classroom your school is given by $g(r) = \pi r^2 \sin \sqrt[3]{r}$ in size and it is circle in shape.
   i. Find the minimum and the maximum size the class could possibly be, if $r = 400$.
   ii. If an average learner occupies an area of 6m², then how many learners can occupy the whole class room?
   iii. Is the class room good in size for an over populated country? Explain.
   iv. What can you conclude to advice construction planners in future?

c) Nelson owns a business that its profit increases by 25% every month, however he wants to make 1 million per month.

## Inverse functions

Within this chapter of study, when dealing with calculus, we will research trigonometric inverse functions within the level of study, however it starts with a question of a Cartesian plane known generally as *y-int of x-inti* [ *f(x)* ].

**Given this figure**

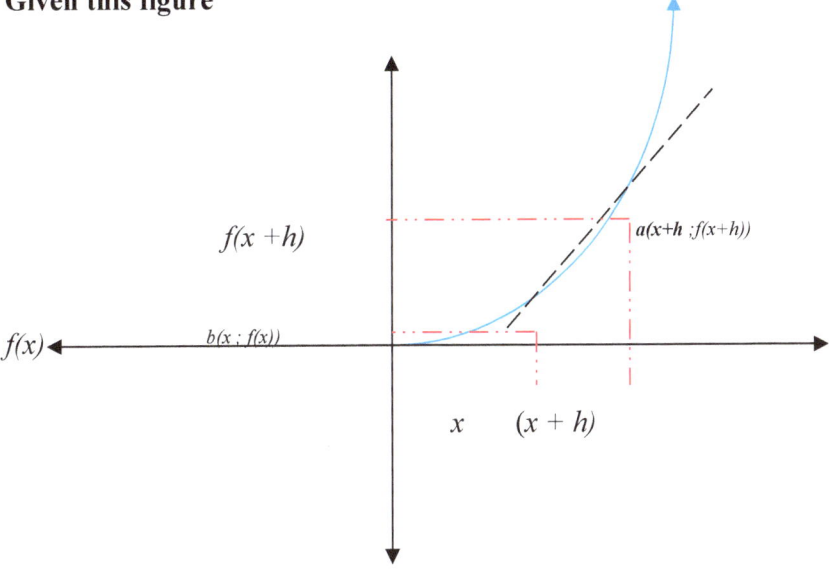

$$\therefore M_{AB} = \frac{y_2 - y_1}{x_2 - x_1}$$

$$= \lim_{h \to 0} \frac{f(x+h) - f(x)}{h}$$

**This is the case within the context of functions and not for inverse functions**

But

**But the inverse function is not philosophy with being denoted by $y = f(x)$**
         **BUT**
**WITH THE PHILOSOPHY OF MATH INTERPRETATION**

$$x \longrightarrow f(y) \longrightarrow f(f(x))$$

*However given this figure*

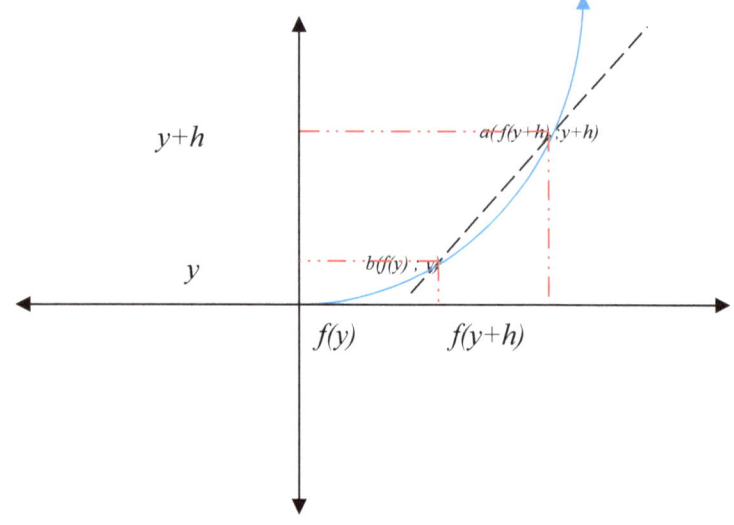

$$M_{ab} = \frac{y_2 - y_1}{x_2 - x_1}$$

$$\therefore = \frac{y + h - y}{f(y + h) - f(y)}$$

$$= \frac{h}{f(y + h) - f(x)}$$

$$\therefore f'(x) = \lim_{h \to 0} \frac{h}{f(y + h) - f(y)}$$

The definition for inverse trigonometric functions

## Inverse function

**Definition states that**

*f(x) is differentiable at x ; if* $\lim_{h \to 0} \frac{h}{f(y + h) - f(y)}$ *it exists. Where f is the original function and f(x) the inverse*

**Definition**

## Skills show

a) Determine the derivative using the definition
   i. $f(x) = \sin^{-1}x$
   ii. $f(x) = \cos^{-1}x$
   iii. $f(x) = \tan^{-1}x$

## Skills show solution

i. $f'(x) = \lim\limits_{h \to 0} \dfrac{h}{f(y+h) - f(y)}$

$= \lim\limits_{h \to 0} \dfrac{h}{\sin(\sin^{-1}x + h) - \sin\sin^{-1}x}$

$= \lim\limits_{h \to 0} \dfrac{h}{\sin\sin^{-1}x . \cos h + \cos\sin^{-1}x . \sin h - \sin\sin^{-1}x}$  **expansion**

$= \lim\limits_{h \to 0} \dfrac{h}{\sin\sin^{-1}x(\cos h - 1) + \cos\sin^{-1}x . \sin h}$  .......**common factor**

at this stage you must introduce $\left(\dfrac{a}{b}\right)_{-1} = \dfrac{b}{a}$

therefore $= \lim\limits_{h \to 0} \left(\dfrac{\sin\sin^{-1}x(\cos h - 1) + \cos\sin^{-1}x . \sin h}{h}\right)_{-1}$

$= \lim\limits_{h \to 0} \left(\dfrac{\sin\sin^{-1}x . (\cos h - 1)}{h} + \dfrac{\cos\sin^{-1}x . \sin h}{h}\right)_{-1}$

$= \left(\sin\sin^{-1}x . \lim\limits_{h \to 0} \dfrac{\cos h - 1}{h} + \cos\sin^{-1}x . \lim\limits_{h \to 0} \dfrac{\sin h}{h}\right)_{-1}$

$= (\cos\sin^{-1}x)_{-1}$

$= \dfrac{1}{\cos\sin^{-1}x}$ ............$\sin^2 x + \cos^2 x = 1$

$= \dfrac{1}{\sqrt{1 - \sin^2\sin^{-1}x}}$ .......$f(f^{-1}(x)) = x$

$= \dfrac{1}{\sqrt{1 - x^2}}$

**The above problem is carefully explained and work done to every detail of the skills and precautions. Hence it is long and detailed to very last law.**

**Considering that this definition is purely from the first principle of differentiation and it is explained as the general concept of the gradient at the specific point. However, the importance of this theory is to further your knowledge within calculus so as to obtain a better understanding of inverse functions.**

ii. If $f(x) = \cos^{-1}x$

$\therefore f'(x) = \lim_{h\to 0} \dfrac{h}{f(y+h) - f(y)}$

$= \lim_{h\to 0} \dfrac{h}{\cos(\cos^{-1}x + h) + \cos\cos^{-1}x}$

$= \lim_{h\to 0} \dfrac{h}{\cos\cos^{-1}x \cdot \cos h - \sin\cos^{-1}x \cdot \sin h + \cos\cos^{-1}x}$

$= \lim_{h\to 0} \dfrac{h}{\cos\cos^{-1}x (\cos h - 1) - \sin\cos^{-1}x \cdot \sin h}$

$= \lim_{h\to 0} \left( \dfrac{\cos\cos^{-1}x \cdot (\cos h - 1) - \sin\cos^{-1}x \cdot \sin h}{h} \right)_{-1}$

$= \lim_{h\to 0} \left( \dfrac{\cos\cos^{-1}x \cdot (\cos h - 1)}{h} - \dfrac{\sin\cos^{-1}x \cdot \sin h}{h} \right)_{-1}$

$= \left( \cos\cos^{-1}x \lim_{h\to 0} \dfrac{\cos h - 1}{h} - \sin\cos^{-1}x \lim_{h\to 0} \dfrac{\sin h}{h} \right)_{-1}$

$= -\dfrac{1}{\sin\cos^{-1}x}$

$= -\dfrac{1}{\sqrt{1-x^2}}$

iii. Try this on your own.

### Verify your skills

a) Find the derivative of the following

i. $f(x) = \sin^{-1}x^2$

ii. $f(x) = \cos^{-1}(\frac{1}{2}x)$

iii. $f(x) = 2\sin^{-1}3x$

iv. $f(x) = 4\cos^{-1}x^2$

b) sketch the functions in a).

c) from the function what can you conclude.

## Conclusion

At the end of this unit of study of calculus you should

- ✓ understand limits
- ✓ master inverse derivatives
- ✓ project the domain and the range of an inverse function
- ✓ understand functions and inverse functions
- ✓ know the gradient at the specific point

## important expressions of calculus

the learner has two choices to choose from . however some learners prefer to use normal expression whereas others prefer to use LEIBNIZ NOTATION which makes calculus a complex study because of its expression

| common easy notation | Leibniz notation |
|---|---|
| $f^1(x) = \ldots\ldots\ldots$ | $\dfrac{dy}{dx} \quad \dfrac{\partial y}{\partial x} \quad \dfrac{\delta y}{\delta x} = \ldots\ldots$ |

<u>*it is up to you which one registers okay in your intelligence box*</u>

## Tangent to a curve

At point A

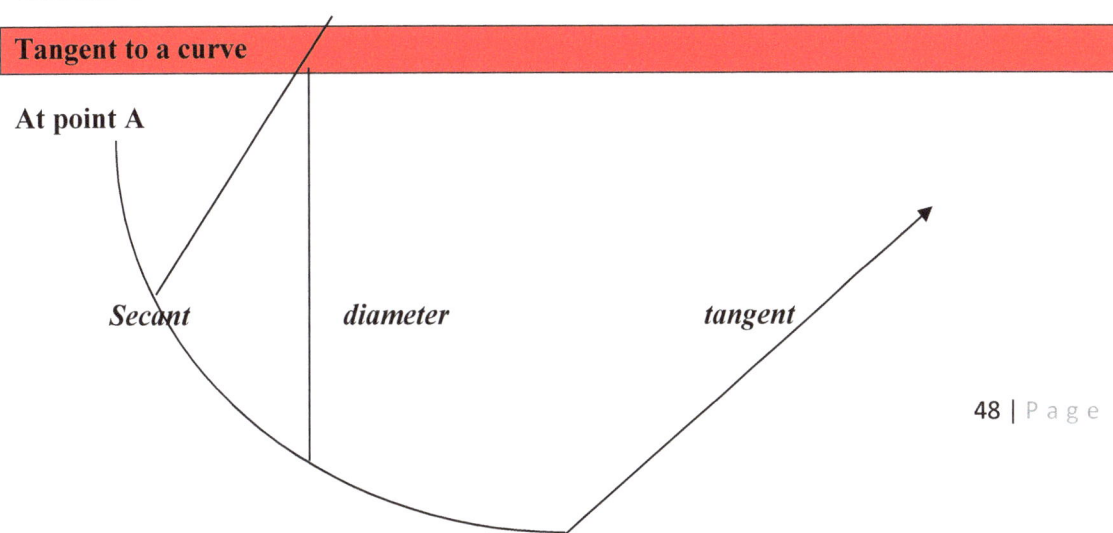

*Secant*     *diameter*     *tangent*

$$y = mx + c$$

A

**Imagine given** $f(x) = ax^2 + bx + c$

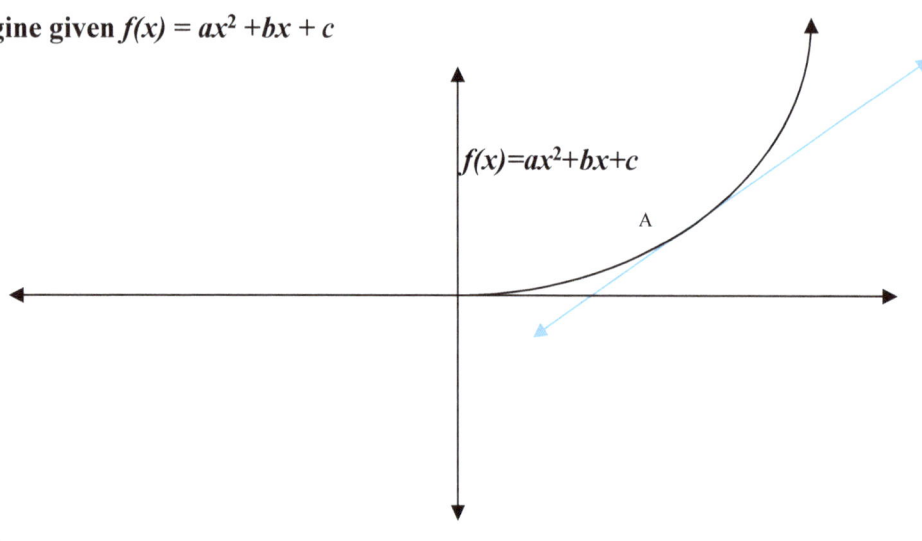

$\therefore f^1(x) = m$

**Which is the gradient at that point**

### Definition

If $f(x)$ at A is differentiable, therefore the line touching point A which is $g(x) = f(x)$ has the gradient defined by $f^1(x)$.

### Skills show

If $f(x) = 3x^2$, find the equation passing through the point $(4\,;\,2)$

$f^1(x) = 3 \cdot 2 \cdot x^{2-1}$
$\quad\quad\, = 6x$

At $x = 4$

$f^1(4) = 6(4)$
$\quad\quad\, = 24$

$\therefore y - y_1 = m(x - x_1)$

$y - 2 = 24(x - 4)$
$y = 24x - 94$

as simple as 1,2,3

## Verify your skills

**Tangent to a curve**
**Work practice**

### Determining the tangent of a curve

a) find the equation to a tangent crossing through the function of $f(x) = \sin x$ at the point $(30, \frac{1}{2})$.

b) Given $f(x) = \cos x$
   i. Show that $f(x) = \cos x$ (the curve) lies on the point $(0, 1)$ and find the equation of the tangent at that point.
   ii. Find the other point on the curve that has the same gradient as the tangent above.

c) Given $f(x) = \sin^{-1} x$
   i. Find the equation of the line / tangent passing through the point where $x = \frac{1}{2}$

## Assignment [ practices on forensics ]

**The was an traffic in Pretoria on the bridge, and this bridge is concave. The curve of that bridge is given by $f(x) = 4\cos^3 x$, on the curve the driver of the truck from the west direction wants to know in calculation that where will the truck move if the brakes are not applied within the curve, provided that he moves 1cm per minute.**

Hint: explore

## Logarithms and exponents

Exponential notation is a mathematical method of expressing extremely large and small numbers . however it is defined as the an expression of a number that question how many times must it be multiplied . e.g. $x^3 = x \cdot x \cdot x$

It was firstly used throughout Greek, Egyptian and Hindu mathematical history , hence today these men are as great mathematicians ever philosophized

- ✓ Leonardo Fibonacci

He helped with the exploration of hindu-arabic mathematical calculation which included the use of exponents.

- ✓ Nicole de Ores me

French mathematician at the Parisian college of Navarre developed (but never published ) "Algorismus proportionum" which introduced notation and computation of frictional elements of exponents.

- ✓ John Napier

Introduction of logarithms

## Exponential limits

*If $f(x) = a^x$*

*Therefore*

$$f'(x) = \lim_{h \to 0} \frac{f(x+h) - f(x)}{h}$$

$$= \lim_{h \to 0} \frac{a^{x+h} - a^x}{h}$$

$$= \lim_{h \to 0} \frac{a^x(a^h - 1)}{h}$$

$$\therefore f'(0) = \lim_{h \to 0} \frac{a^0(a^h - 1)}{h}$$

**Therefore a limit becomes**

$$\lim_{h\to 0} \frac{a^h - 1}{h} = \ln a \quad \text{where } \ln a = \frac{\log a}{\log e} = \log_e a$$

## Skills show

a) Differentiate the following using the limit from the first principle
   i. $f(x) = 2^x$
   ii. $f(x) = 3^{x+2}$
   iii. $f(x) = 6^x + 7$
   iv. $f(x) = 3^x + x^2 + \sin x$ .......brain stretcher............
b) How can you conclude the above.

## Skills show solution

i. $f'(x) = \lim_{h\to 0} \frac{f(x+h) - f(x)}{h}$

$= \lim_{h\to 0} \frac{2^{x+h} - 2^x}{h}$

$= \lim_{h\to 0} \frac{2^x(2^h - 1)}{h}$

$= 2^x \lim_{h\to 0} \frac{2^h - 1}{h}$

$= 2^x \ln 2$

ii. $f'(x) = \lim_{h\to 0} \frac{3^{x+h+2} - 3^{x+2}}{h}$

$= \lim_{h\to 0} \frac{3^{x+2}(3^h - 1)}{h}$

$= 2^{x+2} \lim_{h\to 0} \frac{3^h - 1}{h}$

$= 2^{x+2} \ln 3$

iii. Try this one on your own.

iv. $f(x) = 3^x + x^2 + \sin x$

$f'(x) = \lim_{h\to 0} \frac{f(x+h) - f(x)}{h}$

$= \lim_{h\to 0} \frac{3^{x+h} - (x+h)^2 + \sin(x+h) - 3^x + x^2 + \sin x}{h}$

$= \lim_{h\to 0} \frac{3^x(3^h - 1) + x^2 - 2xh - h^2 + \sin x \cos h + \sin h \cos x - x^2 - \sin x}{h}$

$$= \lim_{h \to 0} \frac{3^x(3^h-1) - h(2x+h) + \sin x(\cos h - 1) + \sin h \cdot \cos x}{h}$$

$$= \lim_{h \to 0} \frac{3^x(3^h-1)}{h} - \frac{h(2x+h)}{h} + \frac{\sin x(\cos h - 1)}{h} + \frac{\sin h \cdot \cos x}{h}$$

$$= 3^x \ln 3 - (2x + (0)) + \sin x \cdot (0) + \cos x \cdot (0)$$

$$= 3^x \ln 3 - 2x + \cos x$$

## Philosophized theory of exponential calculus (10 decimal places)

We will concentrate on the definition of values before 10 decimal places, hence function will be the topic of differential principles within calculus.

## Exponential limit

$$\lim_{h \to 0} \frac{a^h - 1}{h} = \ln a$$

Given the function $f(x) = 2^x$

Therefore $\lim_{h \to 0} \frac{2^h - 1}{h} = \ln 2$

**Considering the data of $h$**

$h_1 = 10^{-1}$      $k = 0.7177346254$

$h_2 = 10^{-2}$      $k = 0.6955550057$

$h_3 = 10^{-3}$      $k = 0.6933874626$

$h_{10} = 10^{-10}$      $k = 0.6931$

$h_{11} = 10^{-11}$      $k = 0.693$

...... $h_{12} = 10^{-12}$      $k = 0$

**The mean/average of the data**

$$\text{Mean}_h = 0.0101010101010101010\ldots\ldots \quad (\text{from } h_1 \text{ to } h_{12})$$

$$f'(x) = \text{mean}_h \lim \frac{a^x(a^h - 1)}{h} \quad \text{to find the average value which is } k$$

definition

remember that the above is calculated using 10 decimal places.

## Reasons why we use 10 decimal places

- ✓ most things begin to be visible at the magnitude of 10 decimal places e.g. carbon dioxide ($CO_2$) where a single atom is microscopic and is not visible by an eye but in nature it appears in moles using avagrado`s constant whereby 1 mol = 6.023 $\times$ $10^{23}$ particles/atoms , hence it be seen .
- ✓ nature present itself visible at 10 decimal places.
- ✓ At the this stage of study this stage of study we only use 10 decimal places at maximum.

However this is theory.

## Conclusion the theorems of calculus by definitions

Provided that this study is commonly expressed as the study of minimum and maximum values of a function , it is then philosophized as the study of understanding the universe and its imperatives such that calculations are carried out to determine the past , present and future predictions of events.

At this level of study you are required to know

- ✓ The use of the derivative
- ✓ Understand the concept of differential calculus
- ✓ Inverse functions of trigonometric calculus
- ✓ The tangent
- ✓ Partial differential calculus as the whole.
- ✓ Master the concept of finding the minimum and the maximum value.
- ✓ The inspection of calculus

As far as this study goes , you are now provided with the fundamentals of it and its definitions by respect of mathematics.

# Applications of Differential calculus

Within this field of calculus we will concentrate on

- ✓ **Cubic functions and quadratic function**
- ✓ **Trigonometric function and its inverses.**
- ✓ **Second derivatives and so on……………..**
- ✓ **Master tangents**
- ✓ **Sketch of graphs**
- ✓ **Rates of change**
- ✓ **Understanding of functions and its inverses**

### Why study this

Thousands of professions need this tool of skill called calculus to interpret the information of events and behaviors.

e.g. engineers

- ✓ chemically            rates of change
- ✓ civil            building a bridge a need to know the minimum quantity

of the cement to build that bridge is determined using this tool of cement.

e.g. if to build a bridge a cement of building is given by the function $f(r) = \pi r^2$
therefore the maximum area per m² is built by the cement of a function of $f^1(r) = 2\pi r$

## Applications

An area with a radius of 4, how much cement of area will it need to build?

$f(4) = 2\pi 4$
      $= 25.13$ units of cement per meter

The above example is based on why certain building collapse and that would be miscalculation of cement mathematically.

### Closure of the derivative

- ✓ derivative is the gradient at the specific point.
- ✓ The process whereby we determine/find the derivative is called partial differentiation or differential calculus.

---

### Tangent

Is a line passing, crossing through and touching the a curve outside

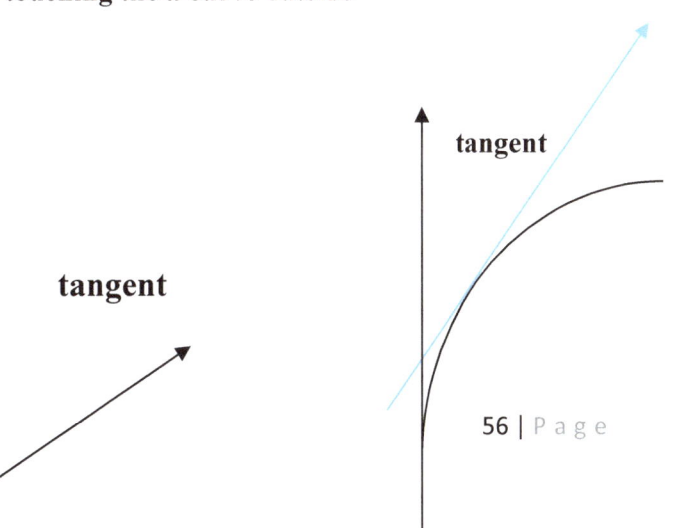

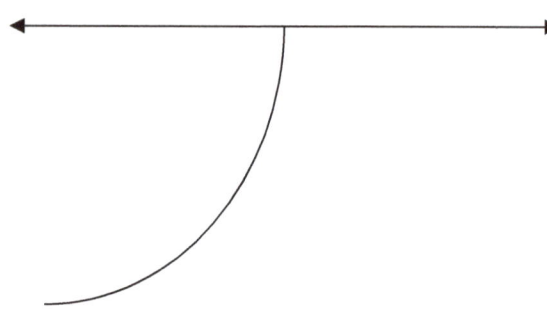

If $f'(x)$ exist then the tangent to the curve where $g(x) = f(x)$ at point A

Is the line passing through A with the gradient given by $f'(x)$.

## Cubic function and tangents

To clearly understand a cubic function we look firstly at

- ✓ the *x*-int and the *y*-int
- ✓ Stationary points
- ✓ Point of inflection

*stationary point  m = 0 maximum*

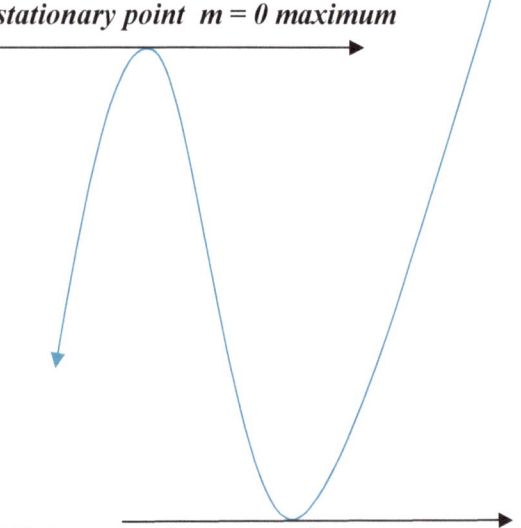

*Minimum stationary m = 0*

the gradient at 180 degrees is 0 which proves that if a $\tan \theta = m$, therefore a *tan 180 = m* which is equal to 0.

## Revision of the gradient

*Positive +*  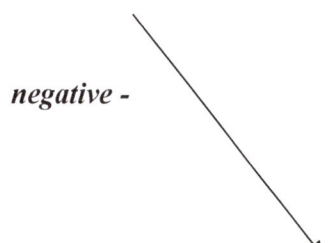  *negative -*

## The structure of a cubic function at critical important points

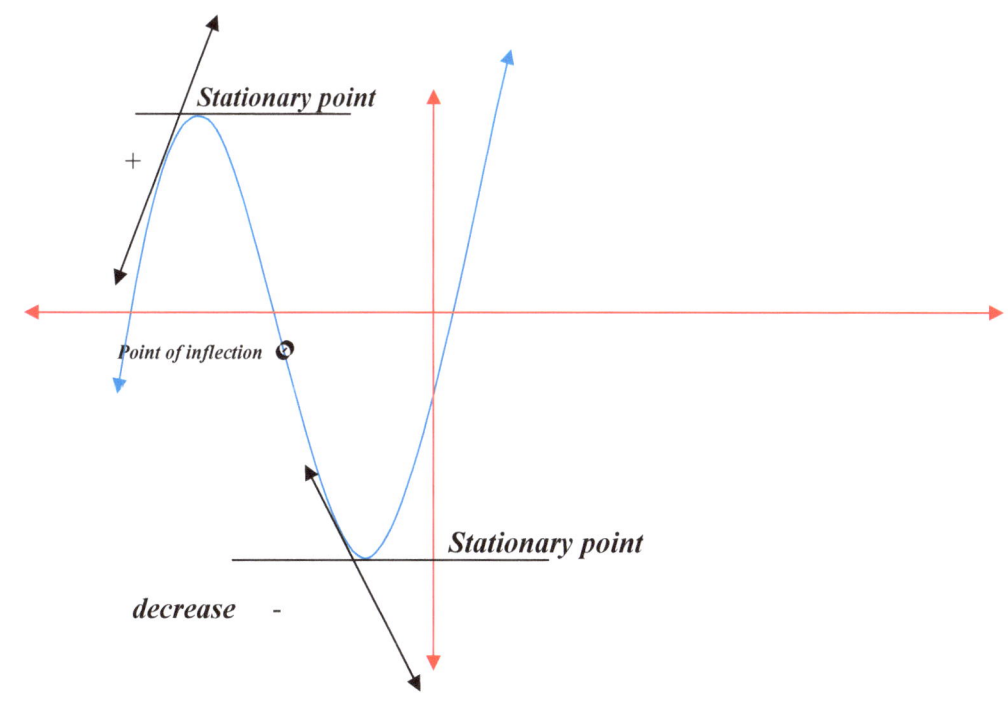

**Analysis of fig: Cubic function**

Given $f(x) = ax^3 + bx^2 + cx + d$

Therefore $f'(x) = 3ax^{3-1} + 2bx^{2-1} + cx^{1-1} = 0$

$3ax^2 + 2bx + c = 0$

$x = \ldots\ldots x_1$ or $x = \ldots\ldots x_2$

$f'(x_1) = \ldots\ldots\ldots$stationary points     Or  $f'(x) = \ldots\ldots\ldots$stationary points

That is minimum and maximum value of the function.

$f''(x) = 6ax + 2b = 0$

$$\frac{6ax}{6a} = \frac{-2b}{6a}$$

Therefore  $x = \dfrac{-b}{3a}$  is a formulae for a point of inflection within a cubic function.

**Skills show A (showing a cubic function)**

a)  Given $f(x) = x^3 - 3x^2 + 4$
  i. Solve for $x$
  ii. Determine the stationary points
  iii. Sketch the graph
  iv. Determine the point of inflection

**Skills show solution**

i.   $f(-1) = 0$
   $\therefore x = -1$
   $x + 1 = 0$

```
            x² - 4x + 4
         ┌─────────────────────
   x + 1 │ x³ – 3x² + 0x + 4
           x³ - x²
           ─────────
           0 - 4x² + 0x
             - 4x² - 4x
             ─────────
                - 4x + 4
                -4x + 4
```

$\therefore (x + 1)(x^2 - 4x + 4) = 0$

$x = -1$ or $x = 2$

ii.                               $f'(x) = 0$

Therefore  $f'(x) = 3x^{3-1} - 3 \cdot 2x^{2-1} = 0$
          $3x^2 - 6x = 0$

$$3x(x-2) = 0$$
$$x = 0 \quad \text{or} \quad x = 2$$
$$f(0) = (0)^3 - 3(0)^2 + 4$$
$$= 4$$
$$f(2) = 2^3 - 3(2)^2 + 4$$
$$= 0$$
$(0, 4)$ and $(2, 0)$

iii.

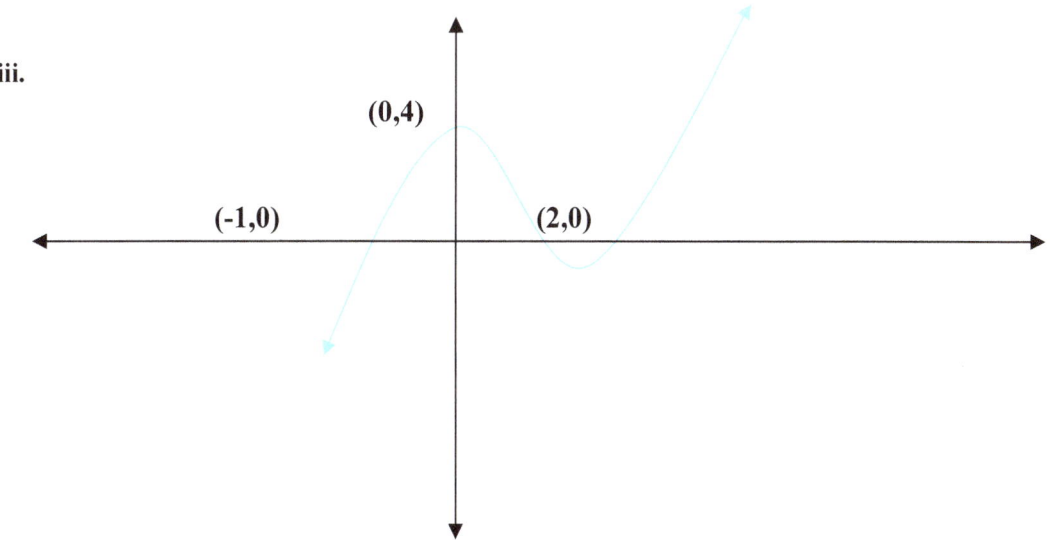

iv.  **Point of inflection**

$$x = \frac{-b}{3a}$$
$$= \frac{-(-3)}{3.1}$$
$$= 1$$

$$f(1) = (1)^3 - 3(1)^2 + 4$$
$$= -2$$

$(1, -2)$ is the point of inflection

## Conclusion

### Approach to a cubic function

- ✓ Finding $x$ intercepts → use reminder theorem for better application
- ✓ Stationary points
- ✓ Point of inflection

## Skills show B [ understanding a cubic function ]

a) Given $f(x) = -x^3 + bx^2 + cx + d$ with turning points at (-1,0) and (1,4).
   i. Find $a$, $b$, $c$ and $d$
   ii. Determine the point of inflection

b) Given $g(x) = ax^3 - 12x^2 + 36x$ with the point of inflection of (4,16)
   i. Find $a$

c) The function $g(x) = x^3 + bx^2 + cx + d$ has the turning points at (0,4) and (2,0)
   i. Prove that $b = -3$, $c = 0$ and $d = 4$.
   ii. Find the equation of the tangent at $x = 2$ and point of inflection.

## Skills show solution

a) i. $a = -1$  ……..given from the equation
turning points (-1,0) and (1,4)

$f'(x) = 0$  …….definition for turning / stationary points

$f'(x) = -3x^2 + 2bx + c = 0$

$x = -1$ or $x = 1$

$x + 1 = 0$ or $x - 1 = 0$

$(x + 1)(x - 1) = 0$

$x^2 - 1 = 0$ ……………raw equation..

$3x^2 - 3x = 0$ ………..multiply by $a$ …..

$2b = 0$ and $c = 3$  from the equation of $f'(x)$

∴ $b = 0$

∴ $f(x) = -x^3 + bx^2 + cx + d$

$$= -x^3 + 0x^2 + 3x + d$$
$$= -x^3 + 3x + d$$

$\therefore f'(x) = -(1)^3 + 3(1) + d = 4$ …….turning points…

$d = 2$

ii. $x = \dfrac{-b}{3a}$ ………point of inflection formulae…..

$= \dfrac{-0}{3(-1)}$

$= 0$

∴ ( 0 , 2 ) is the point of inflection

b) try this one on your own
c) discuss this with a friend.(pairs)

# Trigonometric Calculus (vxps)

**Tangents**

**Consider this figure**

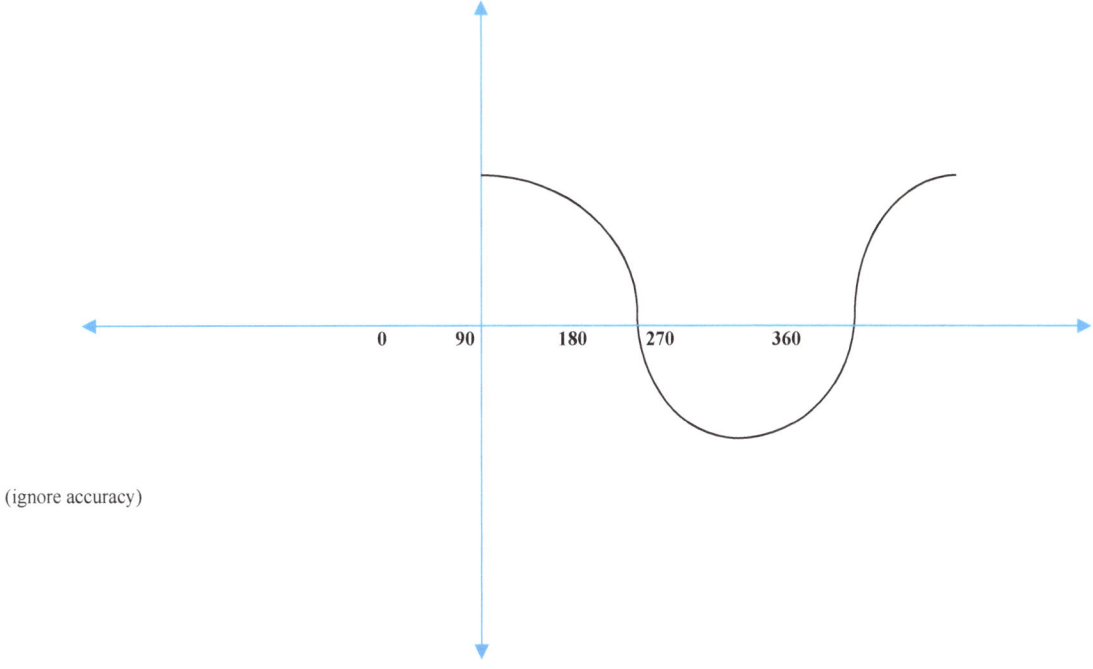

(ignore accuracy)

## Skills show

a) find an equation of the tangent at $x = 60°$

## Skills show solution

a) $f(x) = \cos x$

$\therefore f'(x) = -1.1.1.x^{1-1}\sin x . 1.(\cos x)^{1-\frac{1}{1}}$
$= -\sin x$

$\therefore f'(x) = m$

$-\sin 60 = m$

$m = -\frac{\sqrt{3}}{2}$ ……..gradient at 60 degrees

$y - y_1 = m(x - x_1)$ ….formulae for tangents……

$y - \frac{1}{2} = \left(-\frac{\sqrt{3}}{2}\right)(x - 60)$

$y = -\frac{\sqrt{3}}{2}x + \frac{61\sqrt{3}}{2}$ ………equation of a tangent …………..

## Verify your skills

a) $g(x) = \cos x + 1$
   i. find the equation of a tangent at $x = 30$
   ii. determine if the tangent is decreasing or increasing.
b) Given $f(x) = 3\sin^2 x$
   i. Determine the equation of a tangent at $x = 60$

## Sine function

It is known well that a trig ratio of opposite side over/divided by hypotenuse side is called a sine function, however consider this equation.

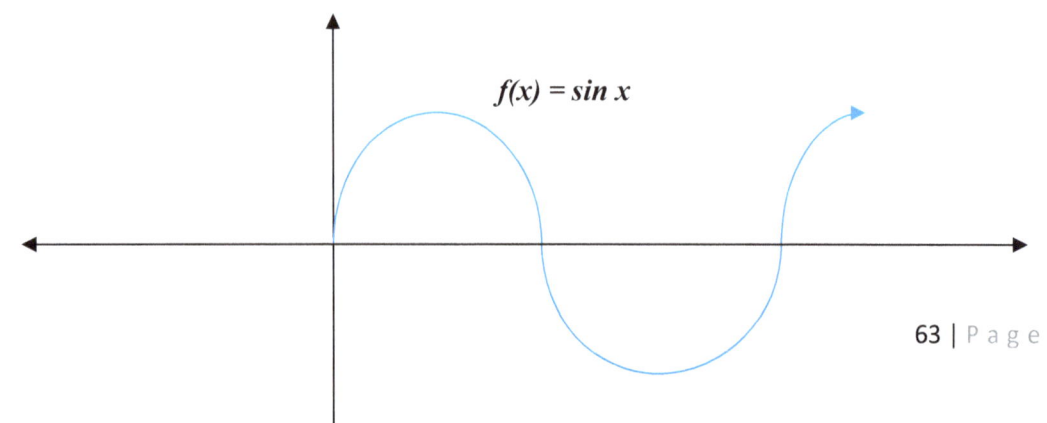

$f(x) = \sin x$

**From the figure above**

$\therefore f'(x) = \cos x = 0$

$x = 90$ ……………..stationary point…….

The information above demonstrates the legitimacy of stationary point

| Looking at the critical points |
|---|

| Skills show |
|---|

Given $f(x) = \sin x$

    i. Determine stationary points [ turning points ]
    ii. Solve for $x$
    iii. Determine the point of inflection
    iv. Give an equation of the tangent passing through the point $x = 45$

| Skills show solution |
|---|

i.    $f(x) = \sin x$

$\therefore f'(x) = abdx^{b-1} \cos ax^b \cdot k(f(x))^{1-\frac{1}{k}}$

$\qquad f'(x) = x^{1-1} \cos x \cdot (\sin x)^{1-\frac{1}{1}}$

$\qquad\qquad = \cos x$

$\therefore f'(x) = \cos x = 0$

$\qquad x = 90$

*general solution because the funtion is 360 degrees*

$x = 90 + k.360 \quad$ or $\quad x = 90+180 + k.360$

$\qquad\qquad\qquad\qquad x = 270 + k.360$

iii.   $f'(x) = \cos x$

$\therefore f^{11}(x) = -1.1.1 \cdot x^{1-1} \sin x \cdot 1.(\cos x)^{1-\frac{1}{1}}$

$\qquad\qquad = -\sin x$

Therefore $x = 0$

The point of inflection is (0,0)
Try doing the rest in pairs.

## Verify your skills

1. Solve for $x$
   a) $x^3 + x^2 - 10x + 8 = 0$
   b) $x^3 + 2x^2 - 19x = 20$
   c) $2x^3 - 7x^2 + 2x = -3$
2. Differentiate from the first principle
   a) $sin^{-1}(7\sqrt{\frac{77}{2}})$
   b) $tan^{-1}(32x^{434})$
   c) $\frac{-2 \pm \sqrt{1 - sin\, x}}{2}$
   d) $\frac{12}{cos\, 2x}$
   e) $(sin^{-1}x - 2)$
   f) $((sin\, 3x)^{-123})sin\, x$

## Assignment

Create your own 25 functions using your intelligence and explore your skills to apprehend and solve them so as to real their characteristics.

## Concluding the study of calculus as a whole

At this stage of knowledge you are now well introduced to the fundamentals of calculus by definitions of being a pure mathematician who understands this tool of study as a great deal to stabilize the concept difficulty. However as far as this study is concerned scientifically the measurement are sometimes excluded form nature and also note that this study determines possibilities and not probabilities as far as the level is concerned.

One person may wonder why this tool should be important. This comes with a conclusion of life challenging impediments e.g. If you were selling fruits and vegetables and you had big ambitions about your business which unfortunately you were not sure of the legitimacy of those dreams. It is then you can use this skill as a tool to determine possibilities.

I then conclude to say is very important that you understand this tool because it is almost done in thousands of professions irrespective of the job to be done.

# *Advancement of analytical Geometry and Measurement*

Centuries ago the Greeks invented a wheel so that the work done of transporting goods can be less. However it has been an advancement of the invention [approximately per decade] that is engineered to the smoothest wheel we see today.

In the scientific world where we look carefully into mediums, it is therefore senseless / non-senseful to achieve a perfect smooth surface. That is if it exists. Therefore

- ✓ We can have friction-less surface.
- ✓ Zero magnitude of vibration between two mediums.

**To this date there is no**

- ✓ Zero friction between two mediums
- ✓ Zero magnitude of vibration between two mediums
- ✓ 100% zero gravity

**Proportionality of the above philosophies goes hand in hand with the following**

- ✓ Shape of the earth
- ✓ Rotation of the planets around the sun
- ✓ Motion of the planets around the sun

**In this unit you will learn**

- ✓ Why anything in motion vibrates
- ✓ The change of concavity of a curve to the smoothest curve
- ✓ π as a risk in calculation
- ✓ A curve and its theory
- ✓ Existence of curve
- ✓ Accuracy and its importance within measurement

**Engineering geometry analytically**

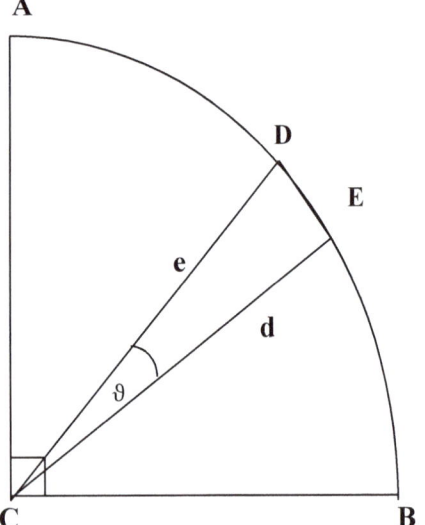

Fig A: Quarter of a circle

$\vartheta = 1°$ ................... per degree unit of measure

$e = d$ ................. radius

therefore

**In triangle DCE**

$c^2 = e^2 + d^2 - 2ed \cos \vartheta$

$e = d$ ………………… *given*

$c^2 = d^2 + d^2 - 2d.d \cos \vartheta$
$c^2 = 2d^2 - 2d^2 \cos \vartheta$
$\therefore c = d\sqrt{2 - 2\cos \vartheta}$
$\quad c = r\sqrt{2 - 2\cos \vartheta}$ ………………… *d is a radius*

## Theorem

$Cr = \text{Ѵ} \, r$    where Cr is the circumference measured in units per degree [ m/degree]

$\text{Ѵ} = \sqrt{2 - 2\cos 1^o}$   is the constant per degree
$r$ is the radius in units e.g. centimeters, meters……

## PI [ π ]

It is known as a ratio of the circumference of a circle to its diameter

$$\frac{\text{Circumference}}{\text{Diameter}} = \pi$$

It is assumed that for all sizes that it is constant throughout.

## History

- Ancient Chinese and the ancient Hebrews seem to have taken π as an equal to 3. A value of it of 3.1605 is implicit in Ahmes Papyrus (c.1650 B.C) of ancient Egypt.
- Archimedes approximated the area of a circle by the method that led to a value of π between $3\frac{1}{7}$ and $3\frac{10}{71}$.

## Conclusion

π is approximated and assumed at different magnitudes to become constant.

## Scientific constant

### Characteristics

- ✓ It is commonly known as a value of measure with theory resulting in disciplines of unit of measure
    e.g. speed is measured by m/s
- ✓ It should obey laws it sets proportional to its theories.

Then you should ask your self this questions

1) Is π theoritical ?
2) Can π set rules that it can obey throughout ?
3) What does π conclude about cirlcles

- At the end of this research you will realize that π is practical and not theoretical meaning that it is not studied and philosophized.
- Rules and laws of mathematics set centuries ago do not circulate around it for comfort of science.

## Theorem of the curve

When

| | |
|---|---|
| $r = 1cm$ | $Cr = 0.017453071 cm/degree$ |
| $r = 2cm$ | $Cr = 0.034906140 cm/degree$ |
| $r = 100cm$ | $Cr = 1.745307108 cm/degree$ |

**Relationship between two points forming a curve**

**A curve has points and within these points in between the gradient changes proportional and indirectly proportional to form a fine curve.**

## Conclusion

$Cr = \mathcal{V} \, r$    where Cr is the circumference measured in units per degree [ m/degree]

$\mathcal{V} = \sqrt{2 - 2\cos 1^o}$  is the constant per degree

$r$ is the radius in units e.g. centimeters, meters……

when the angle decreases the finer the curve becomes .

---

# Techniques of Integration

## Integration by Parts:

This is an intermediate problem designed to help students do trig integrals and recognize when you cannot use natural logs to integrate.

Problem:

Integrate, $\int \dfrac{1}{3x^2 + 2} dx$

Solution:

It is really easy to just write this off as an natural log integral and quickly write the solution as,
$\ln|3x^2 + 2| + C$
BUT THIS IS WRONG! You cannot do a natural log integration with $x^n$ if n is greater than 1.

So you have to do a trig integral.
$$\frac{d}{dx}\left(\frac{1}{a}\tan^{-1}\left(\frac{x}{a}\right)\right) = \frac{1}{x^2 + a^2}$$

First, take out a 1/3 so that you have $x^2$ with no coefficient. $= \frac{1}{3}\int \frac{1}{x^2 + \frac{2}{3}} dx$

Now we have it in the form that we want and we can use solve the integral. It is set up so $a = \sqrt{\frac{2}{3}}$

$$= \frac{1}{3} \cdot \frac{1}{\sqrt{\frac{2}{3}}} \left[\tan^{-1}\left(\frac{x}{\sqrt{\frac{2}{3}}}\right)\right] + C$$

**This is an hard problem designed to help students practice doing integration by parts several times.**

Problem:
Integrate $\int \frac{e^{-4x}}{\sec(3x)} dx$

Solution:

First, you should notice you have to use integration by parts. Notice that 1/sec(3x) is just cos(3x).
$\int u\, dv = uv - \int v\, du$
For this problem, because we have a trig function and a exponential, it does not matter which one you pick to be u and which you pick to be dv.
So, $u = \cos(3x)$ and $dv = e^{-4x} dx$
$du = -3\sin(3x)dx$ and $v = \frac{e^{-4}}{-4}$
now plug those in:
$= \cos(3x) \cdot \left[\frac{e^{-4x}}{-4x}\right] - \int \frac{e^{-4}}{-4} \cdot -3\sin(3x) dx$

here we have to do integration by parts again for $\int \frac{e^{-4x}}{-4} \cdot -3\sin(3x) dx$

so, $u = \sin(3x)$ and $dv = \frac{e^{-4x}}{-4}$
$du = -9\cos(3x)dx$ and $v = \frac{e^{-4x}}{16}$

$$\int e^{-4x} \cdot \cos(3x)dx = \cos(3x) \cdot \left(\frac{e^{-4x}}{-4}\right) - \left\{-3\sin(3x) \cdot \left[\frac{e^{-4x}}{16}\right] - \int \left(\frac{e^{-4x}}{16}\right) \cdot (9\cos(3x)dx)\right\}$$

let $I = \int e^{-4x}\cos(3x)dx$ notice that you have a form of I on both sides of the equation and since I is what we are looking for, we solve for I

$$I = \left[\cos(3x) \cdot \frac{e^{-4x}}{-4}\right] + \left[3\sin(3x) \cdot \frac{e^{-4x}}{16}\right] + \frac{9}{16}I$$ (notice that we pulled the constant (9/16) out of the integral)

Subract (9/16)I from both sides, $\frac{7}{16}I = \cos(3x) \cdot \left[\frac{e^{-4x}}{-4}\right] + 3\sin(3x) \cdot \frac{e^{-4x}}{16}$

Divide both sides by (7/16) to get: $I = \frac{4}{7}\left[\cos(3x) \cdot e^{-4x} + \frac{3}{7}\sin(3x) \cdot e^{-4x}\right]$


**This is an intermediate problem designed to help students identify trig identities and apply them in integration problems. It will also help students practice u-substitutions with sine and cosine**

Problem:

Integrate,

$$\int \frac{\tan^2(x)}{\cos^4(x)}dx$$

Solution:

First, notice that $\frac{1}{\cos^4(x)} = \sec^4(x)$ and separate that into $\sec^2(x) \cdot \sec^2(x)$ so the entire integral becomes $\int \tan^2(x) \cdot \sec^2(x) \cdot \sec^2(x)dx$

Then change one of the $\sec^2(x)$ to $[1 + \tan^2(x)]$, which makes the integral,

$\int \tan^2(x) \cdot [1 + \tan^2(x)] \cdot \sec^2(x)dx$

Then distribute $\tan^2(x)$ over $(1 + \tan^2(x))$, which makes the integral,

$\int [\tan^2(x) + \tan^4(x)] \cdot \sec^2(x)dx$

Then let $u = \tan x, du = \sec^2(x)dx$, so $\int (u^2 + u^4)du$

Then integrate, $\int (u^2 + u^4)du = = (1/3)u^3 + (1/5)u^5$

Then substitute $u = \tan x$ back in, $= (1/3)\tan^3(x) + (1/5)\tan^5(x) + C$


**This is an intermediate problem designed to help students practice partial fractions**

$$\int \frac{\sec^2(x)}{\tan^3(x) + \tan(x)}$$

The hardest part about this problem is figuring out what to do first. Some students might try to rewrite the secant and tangents to simplify the equation. However, the most important thing to do is think about using old integration methods. Like an easy u-substitution.

So, let $u = \tan(x)$ and $du = \sec^2(x)dx$

Now rewrite the integral in terms of u, $\int \dfrac{du}{u^3 + u}$

Now the equation is much simpler and it is easier to see that you should use partial fractions to solve the integral.

Factor out the bottom and put it into "partial fraction form"

$$\int \dfrac{du}{u(u^2+1)}$$

$$= \int \left( \dfrac{A}{u} + \dfrac{Bu+C}{u^2+1} \right)$$

$A(u^2+1) + (Bu+C)u = 1$

now you have to choose values that will cancel out either A or B,C and D
the first choice to cancel out B and C is easy, let u=0
when u=0: $A(1) + 0 = 1 \rightarrow A = 1$
the next one is more difficult to figure out what will make A drop out, many students will notice quickly that -1 will not work and are left to think of a number that will
the number that works this time is i. $i^2 = -1$ and will make A drop out
$A(0) + (Bi+C)i = 1$
$-B + Ci = 1$
$B = -1$
$C = 0$
now go back to the integral and plug in the values you just found

$$= \int \left( \dfrac{1}{u} + \dfrac{-u}{u^2+1} \right) du$$

separate the integral,

$$= \int \dfrac{1}{u} du - \int \dfrac{u}{u^2+1} du$$

$$= \ln|u|$$

for the second integral let v equal the bottom of the fraction, so $dv = 2u\, du$

$$= \dfrac{1}{2} \int \dfrac{dv}{v}$$

$$= \dfrac{1}{2} \ln|v|$$

now plug back in what v and u are equal to, when you do this your final answer will be,

$$\ln|\tan(x)| - \dfrac{1}{2} \ln|\tan^2(x)+1| + C$$

$$= \ln|\tan(x)| - \dfrac{1}{2} \ln|\sec^2(x)| + C$$

**This is a basic problem designed to help students practice finding limits using L'Hopital's rule.**

Problem::

Solve,
$$\lim_{x \to 3}\left(\frac{2x-6}{x^2-9}\right)$$

Solution:

If you just plug in 3 you will get 0/0 which is indeterminate. This is a situation where you can use L'Hopital's rule.

So you take the derivative of the top, divided by the derivative of the bottom:
$$\frac{d}{dx}(2x-6)=2$$
$$\frac{d}{dx}(x^9-9)=2x$$
$$=\frac{2}{2x}$$

now simplify and plug in 3 to get the limit,
$$=\frac{1}{x}$$
$$\lim_{x \to 3}\left(\frac{1}{x}\right)=\frac{1}{3}$$

**This is a basic problem to help students practice solving improper integrals**

Problem:
$$\int_1^\infty \frac{1}{3x}dx$$
To do this, we have to take the limit of this integral as the upper limit goes to infinity, so we write
$$\frac{1}{3}\lim_{b \to \infty}\int_1^b \frac{dx}{x}=\frac{1}{3}\lim_{b \to \infty}[\ln|x|]_1^b=\frac{1}{3}\lim_{b \to \infty}(\ln|b|-\ln|1|)=\frac{1}{3}\lim_{b \to \infty}\ln|b|=\infty$$

## INTEGRATION BY PARTS:

$$\int u\,dv = uv - \int v\,du$$

**PROBLEM: (Hard)**

$$\int \sec^3 x\,dx$$

**SOLUTION:** $\int (\sec^2 x)(\sec x)dx$

$$u = \sec x \qquad\qquad dv = \sec^2 x\,dx$$
$$du = \sec x \tan x \qquad v = \tan x$$

$= \sec x \tan x - \int (\tan x)(\sec x \tan x)dx$

$= \sec x \tan x - \int (\tan^2 x)(\sec x)dx$

$$= \sec x \tan x - \int (\sec^2 x - 1)(\sec x)dx$$

$$= \sec x \tan x - \int (\sec^3 x - \sec x)dx$$

$$= \sec x \tan x - \int (\sec^3 x\, dx + \int \sec x\, dx$$

$$= \sec x \tan x - \int (\sec^3 x\, dx + \int \sec x\, dx$$

$$\int \sec^3 x\, dx = \sec x \tan x - \int \sec^3 x\, dx + \int \sec x \frac{\sec x + \tan x}{\sec x + \tan x} dx$$

$$2\int \sec^3 x\, dx = \sec x \tan x + \int \frac{\sec^2 x + \sec x \tan x}{\sec x + \tan x} dx$$

*Using a "u substitution":*
u=sec x + tan x, du= sec x tan x + sec² x

$$2\int \sec^3 x\, dx = \sec x \tan x + \int \frac{du}{u}$$

$$2\int \sec^3 x\, dx = \sec x \tan x + \ln |\sec x + \tan x|$$

$$\int \sec^3 x\, dx = \frac{sex \tan x + \ln |\sec x \tan x|}{2} + C$$

## U-SUBSTITUTION AND INTEGRATION

**PROBLEM: (hard)**
*Integrate:* $\int \cot x (\ln(\sin x))dx$

**SOLUTION: let u=** $\ln(\sin x)$, **du=** $\frac{1}{\sin x} \cos x\, dx$

**With substitution,** $\int \cot x (\ln(\sin x)dx) = \int u\, du = \frac{u^2}{2}$

$$= \frac{[\ln(\sin x)]^2}{2}$$

**(Easy)**

$$\int \frac{\cos^3 x}{\sqrt{\sin x}} dx$$

**SOLUTION:** $\int \frac{\cos^2 x}{(\sin x)^{1/2}} \cos x\, dx$

$$= \int \frac{1 - \sin^2 x}{(\sin x)^{1/2}} \cos x\, dx$$

**u**= sin x     **du**= cos x dx

$$= \int \frac{1-u^2}{u^{1/2}} du$$

$$= \int \frac{1}{u^{1/2}} du - \int u^{3/2} du = \int u^{-1/2} du - \int u^{3/2} du$$

$$= 2u^{1/2} - \frac{2u^{5/2}}{5}$$

$$= 2\sin^{1/2} x - \frac{2}{5}\sin^{5/2} x$$

(Medium)

$$\int \sec^4(6x)\tan^3(6x)\,dx$$

**SOLUTION:** $\int \sec^4(6x)\tan^3(6x)\,dx$

**Pull out a sec²x**

$$= \int \sec^2(6x)\tan^3(6x)(\sec^2(6x))\,dx$$

$$= \int [1-\tan^2(6x)][\tan^3(6x)](\sec^2(6x))\,dx$$

**Let u**= $\tan(6x)$, **du**=$\sec^2(6x)dx$

$$= \int (1-u^2)(u^3)\,du$$

$$= \int (u^3 - u^5)\,du$$

$$= \int u^3 du - \int u^5 du$$

$$= \frac{u^4}{4} - \frac{u^6}{6}$$

$$= \frac{1}{4}\tan^4(6x) - \frac{1}{6}\tan^6(6x)$$

## *PARTIAL FRACTIONS:*

**PROBLEM: (Medium)**

$$\int \frac{\cos x}{\sin x(\sin x - 1)}\,dx$$

**SOLUTION: Let** $u = \sin x$, $du = \cos x\,dx$

$$= \int \frac{du}{u(u-1)}$$

$$= \frac{1}{u(u-1)} = \frac{A}{u} + \frac{B}{u-1}$$

$$\to 1 = A(u-1) + B(u)$$

$u = 1, B = 1$

$u = 0, A = -1$

$= \int \dfrac{du}{u-1} - \int \dfrac{du}{u}$

$= \ln|u-1| - \ln|u|$

$= \ln\left|\dfrac{u-1}{u}\right|$

$= \ln\left|\dfrac{\sin x - 1}{\sin x}\right|$

## INTEGRATION BY PARTS

**PROBLEM: (Hard)**

$\int e^{3x} \cos\dfrac{1}{3}x\, dx$

**SOLUTION:** $\int e^{3x} \cos\dfrac{1}{3}x\, dx$

$u = e^{3x}$ $\qquad dv = \cos\dfrac{1}{3}x\, dx$

$du = 3e^{3x}$ $\qquad v = 3\sin\dfrac{1}{3}x$

$* uv - \int v\, du$

$= 3e^{3x} \sin\dfrac{1}{3}x - \int 9e^{3x} \sin\dfrac{1}{3}x\, dx$

$u = 9e^{3x}$ $\qquad dv = \sin\dfrac{1}{3}x\, dx$

$du = 27e^{3x}$ $\qquad v = -3\cos\dfrac{1}{3}x$

$= 3e^{3x} \sin\dfrac{1}{3}x - 27e^{3x} \cos\dfrac{1}{3}x + 81\int e^{3x} \cos\dfrac{1}{3}x\, dx$

$\int e^{3x} \cos\dfrac{1}{3}x\, dx = 3e^{3x} \sin\dfrac{1}{3}x - 27e^{3x} \cos\dfrac{1}{3}x + 81\int e^{3x} \cos\dfrac{1}{3}x\, dx$

$-80 \int e^{3x} \cos\dfrac{1}{3}x\, dx = 3e^{3x} \sin\dfrac{1}{3}x - 27e^{3x} \cos\dfrac{1}{3}x$

$\rightarrow \int e^{3x} \cos\dfrac{1}{3}x\, dx = -\dfrac{3e^{3x} \sin\dfrac{1}{3}x - 27e^{3x} \cos\dfrac{1}{3}x}{80}$

**Improper Integrals**

Solve $\int_1^\infty \frac{dx}{x^5}$. Remember to use those limits of b!

$$\int_1^\infty \frac{dx}{x^5} = \lim_{b \to \infty} \int_1^b \frac{dx}{x^5}$$

$$= \lim_{b \to \infty} \int_1^b x^{-5} dx = \lim_{b \to \infty} [\frac{-1}{4} x^{-4}]_1^b$$

$$= \lim_{b \to \infty} [\frac{-1}{4} b^{-4} + \frac{1}{4}(1)^{-4}]_1^b$$

$$= \lim_{b \to \infty} [\frac{-1}{4b^4} + \frac{1}{4}]_1^b = [0 + \frac{1}{4}]$$

$$= \frac{1}{4}$$

### 1 Deutsches Bargeldwachstum

Herr Geldkostenzähler, a former Reichsbank employee of the early 1920's, graphs German inflation on the remains of his tattered bed spread (which is his only possession) in blood. It is in the shape of ln(x) from x=1 meter to x=3 meters.

If rope to hang yourself with costs 27 billion marks per meter at 3:00 PM, how much would suicide cost Herr Geldkostenzähler if he bought some hanging rope as long as his blood streak? You will have to use trigonometric substitution (hint: let x=cotθ) and integration by parts.

*First find the arc length of the blood streak (ln x):*

$$s = \int_1^3 (1 + f'(x)^2)^{1/2} dx$$

$$= \int_1^3 (1 + \frac{1}{x^2})^{1/2} dx$$

Let $x = \cot(\theta) \Rightarrow dx = -\csc^2(\theta) d\theta$

$$\Rightarrow \int (1 + \frac{1}{x^2})^{1/2} dx = -\int [1 + \tan^2 \theta]^{1/2} \csc^2(\theta) d\theta$$

$$= -\int \sec(\theta) \csc^2(\theta) d\theta$$

*Integration by parts:*

$$\int u\,dv = uv - \int v\,du$$

Let $u = \sec\theta \Rightarrow du = \tan\theta \sec\theta\, d\theta$

Let $dv = \csc^2 \theta\, d\theta \Rightarrow v = -\cot\theta$

$$\Rightarrow -\int \sec(\theta) \csc^2(\theta) d\theta = \sec(\theta)\cot\theta - \int \cot(\theta)\tan(\theta)\sec(\theta) d\theta$$

$$= \sec\theta\tan - \int \sec\theta\, d\theta$$

*The rest is common sense!*

$$\int \sec\theta d\theta = \int \sec\theta [\frac{\sec\theta + \tan\theta}{\sec\theta + \tan\theta}]d\theta$$

$$= \int \frac{\sec^2\theta + \sec\theta\tan\theta}{\sec\theta + \tan\theta} d\theta$$

Let $u = \sec\theta + \tan\theta \Rightarrow du = (\sec\theta\tan\theta + \sec^2\theta)d\theta$

$$= \int \frac{du}{u} = \ln(u) = \ln(\sec\theta = \tan\theta)$$

*…Or at least sense as common as it could be under the circumstances…*

$$\Rightarrow \sec\theta\tan\theta - \int \sec\theta d\theta$$

$$= \sec\theta\tan\theta - \ln(\sec\theta + \tan\theta)$$

$$= \left[(\frac{\sqrt{x^2+1}}{x})x - \ln(\frac{\sqrt{x^2+1}}{x} + \frac{1}{x})\right]_1^3$$

$$= \left[\sqrt{10} - \ln(\frac{\sqrt{10}}{3} + \frac{1}{3})\right] - \left[\sqrt{2} - \ln(\frac{\sqrt{2}}{1} + 1)\right]$$

$$= 2.301987535$$

$$\approx 2.3 \ meters$$

Since rope costs 27 billion marks per meter, Herr Geldkostenzähler will have to cough up about **62.1 billion marks** to leave this world behind, assuming that the price of rope hasn't risen in the time it takes him to figure out how much he owes.

2
3 **Improper Integrals**

Solve $\int_0^1 x\ln x \ dx$

*Use integration by parts:*

$$\int_0^1 x\ln x\, dx = \lim_{b\to 0}\int_b^1 x\ln x\, dx$$

Let $u = \ln x$ and $dv = x\, dx$

$$du = \frac{dx}{x} \quad v = \frac{1}{2}x^2$$

$$(\ln x)(\frac{1}{2}x^2) - \int (\frac{1}{2}x^2)(\frac{1}{x})dx$$

$$= \lim_{b\to 0}\left[(\ln x)(\frac{1}{2}x^2) - \frac{x^2}{4}\right]_b^1$$

$$= \lim_{b\to 0}\left[\frac{1}{2}(1)^2\ln(1) - \frac{(1)^2}{4} - \frac{1}{2}(b)^2\ln(b) - \frac{(b)^2}{4}\right]_b^1$$

$$= \lim_{b\to 0}\left[0 - \frac{1}{4} - \frac{1}{2}(b)^2\ln(b) - \frac{(b)^2}{4}\right]_b^1$$

The $\lim_{b\to 0}(b^2\ln b)$ is indeterminate in this form, so rewrite it as $\lim_{b\to 0}(\frac{\ln b}{b^2})$ and use L'Hôpital's Rule.

$$\lim_{b\to 0}(\frac{\ln b}{b^2}) = \lim_{b\to 0}(\frac{1/b}{-2/b^2}) = \lim_{b\to 0}(\frac{-b^2}{2}) = 0$$

Therefore:

$$\int_0^1 x\ln x\, dx = \lim_{b\to 0}\left[0 - 1/4 - 0 - 0\right] = \frac{1}{4}$$

4

5    **Surface Area**

The arc of the curve of $y = \ln x$ lying in the fourth quadrant is revolved around the y-axis. Find the surface area generated by the created volume.

$$ds = \left[1 + \left(\frac{dx}{dy}\right)^2\right]^{\frac{1}{2}} dy$$

$$dS = 2\pi x\, ds = 2\pi x\left[1 + \left(\frac{dx}{dy}\right)^2\right]^{\frac{1}{2}} dy$$

$y = \ln x$

$$\Rightarrow \frac{dy}{dx} = \frac{1}{x} \Rightarrow \frac{dx}{dy} = x$$

$$dy = \frac{dx}{x}$$

*Upon substitution:*

$$dS = 2\pi x \left[1 + \left(\frac{dx}{dy}\right)^2\right]^{\frac{1}{2}} dy$$

$$= 2\pi x \left[1 + x^2\right]^{\frac{1}{2}} \left(\frac{dx}{x}\right)$$

$$= 2\pi \left[1 + x^2\right]^{\frac{1}{2}} dx$$

$$\Rightarrow S = 2\pi \int_0^1 \left[1 + x^2\right]^{\frac{1}{2}} dx$$

Let $x = \tan\theta$ and $dx = \sec^2\theta \, d\theta$

$$\Rightarrow S = 2\pi \int \left[1 + \tan^2\theta\right]^{\frac{1}{2}} \sec^2\theta \, d\theta$$

$$= 2\pi \int (\sec^2\theta)^{\frac{1}{2}} \sec^2\theta \, d\theta = 2\pi \int \sec^3\theta \, d\theta$$

*Integrate by parts:*

Let $u = \sec\theta$     $dv = \sec^2\theta \, d\theta$
$du = \sec\theta \, d\theta$     $v = \tan\theta$

$$2\pi \int \sec^3\theta \, d\theta$$

$$= 2\pi \left[\tan\theta \sec\theta - \int \tan^2\theta \sec\theta \, d\theta\right]$$

$$= 2\pi \left[\tan\theta \sec\theta - \int (\sec^2\theta - 1)\sec\theta \, d\theta\right]$$

$$= 2\pi \left[\tan\theta \sec\theta - \int (\sec^3\theta - \sec\theta) d\theta\right]$$

$$= 2\pi \left[\frac{\tan\theta \sec\theta + \int \sec\theta \, d\theta}{2}\right]$$

$$= 2\pi \left[\frac{\tan\theta \sec\theta + \ln(\tan\theta + \sec\theta)}{2}\right]$$

$$= \pi \left[\tan\theta \sec\theta + \ln[\tan\theta + \sec\theta]\right]$$

$$\therefore S = \pi \left[x\sqrt{x^2 + 1} + \ln\left(x + \sqrt{x^2 + 1}\right)\right]_0^1$$

$$\Rightarrow S = \pi \left[\sqrt{2} + \ln\left(\sqrt{2} + 1\right)\right]$$

### 6    Integral or Something

Find the anti-derivative of $\int \dfrac{\left[\dfrac{8\sin\theta}{\cos^3\theta} + \dfrac{2}{\cos^2\theta}\right] d\theta}{\tan^2\theta - 1}$. You will have to use **partial fractions**.

*First simplify the integral to make it less nasty, then use a substitution followed by partial fractions to seal the deal.*

$$\int \frac{\left[\dfrac{8\sin\theta}{\cos^3\theta} + \dfrac{2}{\cos^2\theta}\right] d\theta}{\tan^2\theta - 1}$$

$$= \int \frac{8\tan\theta\sec^2\theta + 2\sec^2\theta}{\tan^2\theta - 1} d\theta$$

$$= \int \frac{\sec^2\theta \, [8\tan\theta + 2]}{\tan^2\theta - 1} d\theta$$

Substitution:

$u = \tan\theta \; ; \; du = \sec^2\theta$

$$\Rightarrow \int \frac{\sec^2\theta \, [8(\tan\theta) + 2]}{\tan^2\theta - 1} d\theta = \int \frac{8u + 2}{u^2 - 1} du$$

$$\int \frac{8u+2}{u^2-1} du = \int \frac{8u+2}{(u+1)(u-1)} dv$$

Solve by partial fractions:

$$\frac{8u+2}{(u+1)(u-1)} = \frac{A}{u+1} + \frac{B}{u-1}$$

$\Rightarrow 8u + 2 = A(u-1) + B(u+1)$

$\Rightarrow A = 3 \; ; \; B = 5$

$$\Rightarrow \frac{8u+2}{(u+1)(u-1)} = \frac{3}{u+1} + \frac{5}{u-1}$$

Therefore:

$$\int \frac{8u+2}{(u+1)(u-1)} du = \int \left[\frac{3}{u+1} + \frac{5}{u-1}\right] du =$$

$$= \int \frac{3}{u+1} du + \int \frac{5}{u-1} du$$

$$= 3\ln|u+1| + 5\ln|u-1| + C$$

*Don't forget to put add a +C and revert back to your theta variables.*

$3\ln|u+1| + 5\ln|u-1| + C$

$= 3\ln|\tan\theta + 1| + 5\ln|\tan\theta - 1| + C$

This is an advanced-level problem. Students need to be familiar with the concept of **integration by parts**, and will also be required to derive the integral of tanx, and use a u-substitution. This problem could be used in a review that covers several topics, after students have learned integration by parts.

$$\int x\sec^2 x \, dx = ?$$

Solution: For this problem we will begin by using integration by parts. We will then have to derive the integral of tanx and use a u-substitution in order to find our solution.

For integration by parts, remember: $\int u\,dv = uv - \int v\,du$

The goal is to ensure that vdu is simpler than udv. So, we set u = x and dv = sec²xdx. Then du=dx, and v=tanx. So,

$$\int x\sec^2 x\,dx = \int u\,dv = uv - \int v\,du$$
$$= x\tan x - \int \tan x\,dx$$

Now, we need to figure out what $\int \tan x\,dx$ is:

$$\int \tan x\,dx = \int \frac{\sin x}{\cos x}\,dx$$

let u = cosx, du = -sinxdx

$$= \int -\frac{du}{u} = -\ln|u| = \ln\left|\frac{1}{u}\right| = \ln\left|\frac{1}{\cos x}\right| = \ln|\sec x|$$

Therefore,

$$\int x\sec^2 x\,dx = x\tan x - \int \tan x\,dx = x\tan x - \ln|\sec x| + C$$

This is an advanced-level problem. Students need to be able to do **partial fractions**, as well as u-substitutions. This could be used as a more advanced problem after students are familiar with these concepts.

Integrate: $\int \dfrac{3x+2}{2x^2+3x+1}\,dx$

Solution:
To solve this problem, we will first use integration by parts, then use two u-substitutions.

First, you need to factor the denominator:

$$\int \frac{3x+2}{(2x+1)(x+1)}\,dx$$

We can now use partial fractions:

$$\frac{3x+1}{(2x+1)(x+1)} = \frac{A}{2x+1} + \frac{B}{x+1}$$

Eliminate fractions:
$3x+2 = A(x+1) + B(2x+1)$
$= Ax + A + 2Bx + B$
$= x(A+2B) + (A+B)$

Therefore, A+2B = 3 and A+B = 2

$$A=B=1$$

$$\int \frac{3x+2}{(2x+1)(x+1)}\,dx = \int\left(\frac{1}{2x+1} + \frac{1}{x+1}\right)dx = \int\frac{1}{2x+1}\,dx + \int\frac{1}{x+1}\,dx$$

For the first, let u = 2x+1, du = 2dx:  $\frac{1}{2}\int\frac{du}{u} = \frac{1}{2}\ln|u| = \frac{1}{2}\ln|2x+1|$

For the second part let u = x+1, du = dx: $\int\frac{du}{u} = \ln|u| = \ln|x+1|$

Putting it together, our final answer is: $\frac{1}{2}\ln|2x+1| + \ln|x+1| + C$

**Evaluate** $\int_0^1 \frac{2x^2}{7x-4}dx$ .

***Fix: Do as improper integral.***
Check to see if function is continuous over the interval (0, 1).

Function is discontinuous when the denominator is equal to zero:

7x – 4 = 0

$x = \frac{4}{7}$    Integral is undefined - Fundamental Theorem of Calculus doesn't apply because of discontinuity at $x = \frac{4}{7}$.

**Evaluate** $\int (x^3 3^x\ dx)$.

Solution:        Tabular Method:

$$\int (x^3 3^x\ dx) = \frac{3^x x^3}{\ln 3} - \frac{3^x x^2}{(\ln 3)^2} + \frac{3^x 6x}{(\ln 3)^3} - \frac{3^x(6)}{(\ln 3)^4} + C$$

| | | |
|---|---|---|
| + | $x^3$ | $3^x$ |
| − | $3x^2$ | $\dfrac{3^x}{(\ln 3)}$ |
| + | $6x$ | $\dfrac{3^x}{(\ln 3)^2}$ |
| − | $6$ | $\dfrac{3^x}{(\ln 3)^3}$ |
| + | $0$ | $\dfrac{3^x}{(\ln 3)^4}$ |

L'Hopital's Rule, **Evaluate** $\lim_{x\to 0^-} x^x$.

Solution:    $y = \lim_{x\to 0^-} x^x$

ln (y) = $\lim_{x\to 0^-} \ln(x^x)$

ln (y) = $\lim_{x\to 0^-} [x\ln(x)]$

ln (y) = $\lim_{x\to 0^-} \frac{\ln(x)}{\frac{1}{x}} = \frac{\infty}{\infty}$   Put what's inside the limit in indeterminate form so that we can use L'Hopital.

L'Hopital: $\ln(y) = \lim\limits_{x \to 0^-} \dfrac{\dfrac{d}{dx}\ln(x)}{\dfrac{d}{dx}\left(\dfrac{1}{x}\right)} = \lim\limits_{x \to 0^-} \dfrac{\dfrac{1}{x}}{-x^{-2}} = \lim\limits_{x \to 0^-} -x = 0$

$y = e^0$
$y = 1$

$\lim\limits_{x \to 0^-} x^x = 1$

**Partial Fractions, u-substitution, trig-substitution, and basic triangle trig Chap 8: Evaluate**
$\int \dfrac{7x+4\ dx}{x(3x+2)^2}$.

Solution: Split the integrand into partial fractions: $\dfrac{7x+4}{x(3x+2)^2} = \dfrac{A}{x} + \dfrac{B}{3x+2} + \dfrac{C}{(3x+2)^2}$

$A(3x+2)^2 + B(x)(3x+2) + C(x) = 7x + 4$

$(9x^2 + 12x + 4)A + (3x^2+2x)B + Cx = 7x + 4$

$9Ax^2 + 3Bx^2 + 12Ax + 2Bx + Cx + 4A = 7x + 4$

$(9A + 3B)x^2 + (12A + 2B + C)x + 4A = 7x + 4$

Coefficients on the left have to match up with the coefficients on the right, so:

4A = 4
A = 1

$9x^2A + 3x^2B = 0x^2$
        9A + 3B = 0
        9(1)+3B=0
        B = -3

12xA + 2xB + Cx = 7x
12A + 2B + C = 7
12 – 6 + C =7
C = 1

Substitute A, B, and C back into the partial fractions: $\dfrac{7x+4}{x(3x+2)^2} = \dfrac{1}{x} + \dfrac{-3}{3x+2} + \dfrac{1}{(3x+2)^2}$

$$\int \dfrac{7x+4\ dx}{x(3x+2)^2} = \int \dfrac{1}{x}dx + \int \dfrac{-3}{3x+2}dx + \int \dfrac{1}{(3x+2)^2}dx$$

$$\int \frac{1}{x} dx = \ln|x| + C$$

U Substitution: $\int \frac{-3}{3x+2} dx \Rightarrow u = 3x+2 \quad du = 3\, dx$

$$\int \frac{-3}{3x+2} dx = \int \frac{-du}{u} = -\ln|3x+2| + C$$

Trig Substitution: $\int \frac{1}{(3x+2)^2} dx \Rightarrow x = \frac{2}{3}\tan^2\theta \quad dx = \frac{4}{3}\tan\theta \sec^2\theta\, d\theta$

$$\int \frac{1}{(3x+2)^2} dx = \int \frac{\frac{4}{3}\tan\theta\sec^2\theta}{\left((3)\frac{2}{3}\tan^2\theta + 2\right)^2} d\theta = \int \frac{\frac{4}{3}\tan\theta\sec^2\theta}{2^2(\tan^2\theta+1)^2} d\theta = \frac{1}{3}\int \frac{\tan\theta\sec^2\theta}{\sec^4\theta} d\theta = \frac{1}{3}\int \tan\theta\cos^2\theta\, d\theta$$

V Substitution: $= \frac{1}{3}\int \tan\theta\cos^2\theta\, d\theta = \frac{1}{3}\int \sin\theta\cos\theta\, d\theta \Rightarrow v = \sin\theta \quad dv = \cos\theta\, d\theta$

$$= \frac{1}{3}\int v\, dv = \frac{1}{6}v^2 + C = \frac{\sin^2\theta}{6} + C = \frac{3x}{6(3x+2)} + C = \frac{x}{6x+4} + C$$

$x = \frac{2}{3}\tan^2\theta$

$\tan^{-1}\theta = \sqrt{\frac{3}{2}(x)}$

$\sin\theta = \frac{\sqrt{3x}}{\sqrt{3x+2}}$

$$\int \frac{7x+4\, dx}{x(3x+2)^2} = \ln|x| - \ln|3x+2| + \frac{x}{6x+4} + C = \ln\left|\frac{x}{3x+2}\right| + \frac{x}{6x+4} + C$$

**Cauchy Mean Value Theorem:** pg 615, Salas

**L'Hospital's Rule:**

**L'Hospital's rule** Find $\lim\limits_{x \to \infty} \frac{7x^4 + 5x^2 + 3x + 1}{3x^4 + 7x}$.

Solution: Since both numerator and denominator tend toward infinity as $x \to \infty$, we have the indeterminate form $\infty/\infty$. L'Hospital's rule does apply to this form. We begin by taking the limit as $x \to \infty$ of the derivative of the top over the derivative of the bottom. In doing so, we see that we will have to use L'Hospital's rule repeatedly. Eventually, we arrive at the limit of a constant:

$$\lim_{x \to \infty}\frac{7x^4+5x^2+3x+1}{3x^4+7x} = \lim_{x \to \infty}\frac{28x^3+10x+3}{12x^3+7} = \lim_{x \to \infty}\frac{84x^2+10}{36x^2} = \lim_{x \to \infty}\frac{168x}{72x} = \lim_{x \to \infty}\frac{168}{72} = \frac{168}{72}$$
$$= 2.33.$$

1. Clearly explaining the relationship between L'Hospital's rule and horizontal asymptotes (if they exist) for rational functions.

Using **L'Hopital's Rule** to evaluate limits of function $\lim_{x \to 0}\dfrac{2-\sqrt{x^2+4}}{x^2}$ that are in indeterminate form $0/0$.

Solution: $\lim_{x \to 0}\dfrac{2-\sqrt{x^2+4}}{x^2} = \lim_{x \to 0}\dfrac{-\frac{1}{2}(x^2+4)^{-\frac{1}{2}}(2x)}{2x} = \lim_{x \to 0}\dfrac{-(x^2+4)^{-\frac{1}{2}}-(-x)(-\frac{1}{2})(x^2+4)^{-\frac{3}{2}}}{2}$

$= \lim_{x \to 0}\dfrac{1}{2}\left(-\dfrac{1}{\sqrt{x^2+4}} - \dfrac{x}{2\sqrt{(x^2+4)^3}}\right) = \dfrac{1}{2}(-\dfrac{1}{2}-0) = -\dfrac{1}{4}.$

Does the integral $\int_2^\infty \dfrac{1}{x^3+1}dx$ converge?

(using the Limit Comparison Test to evaluate if a function converge or diverge)

Solution: By definition, $\int_2^\infty \dfrac{1}{x^3+1}dx = \lim_{N \to \infty}\int_2^N \dfrac{1}{x^3+1}dx$

We cannot evaluate the latter integral, there is no simple formula for the $\dfrac{1}{x^3+1}$.

Since $\dfrac{1}{x^3+1} > 0$ on $[2, \infty]$ and $\dfrac{1}{x^3+1} < \dfrac{1}{x^3}$, $\int_2^\infty \dfrac{1}{x^3+1}dx < \int_2^\infty \dfrac{1}{x^3}$.

$\int_2^\infty \dfrac{1}{x^3}$

$= \lim_{N \to \infty}\int_2^N \dfrac{1}{x^3}$

$= \lim_{N \to \infty}\left[-\dfrac{1}{2}x^{-2}\right]_2^N$

$= \lim_{N \to \infty}\left[-\dfrac{1}{2N^2}-(-\dfrac{1}{2(2^2)})\right]$

$= 0 + \dfrac{1}{8}$

$= \dfrac{1}{8} < \infty$

Therefore, the given infinite integral $\int_{2}^{\infty} \frac{1}{x^3+1} dx$ converges.

Find the volume of the solid when the region bounded by $f(x) = \sqrt{x \sin x + 2}$ and $g(x) = x^2 + 1$ is rotated around the x-axis. (using Washer method to find the volume of a solid and use the short cut of Integration by Parts, Tabular Method)

Solution:
$\sqrt{x \sin x + 2} = x^2 + 1$

intercept on x: $\pm .7696426$

$V = \pi \int_{-.7696426}^{.7696426} \left[ \sqrt{x\sin x + 2}^2 - (x^2 + 1)^2 \right] dx$

$= \pi \int_{-.7696426}^{.7696426} \left[ x \sin x - x^4 - 2x^2 + 1 \right] dx V$

$= \pi \int x \sin x \, dx - \pi \int (x^4 + 2x^2 - 1) dx$

$= \pi \int x \sin x \, dx - \pi [\frac{1}{5} x^5 + \frac{2}{3} x^3 - x]$

| u | dv/dx | Tabular method |
|---|---|---|
| + x | sin x | $\int x \sin x \, dx = -x \cos x + \sin x$ |
| - 1 | -cos x | |
| + 0 | -sin x | |

$V = \left[ -\pi x \cos x + \pi \sin x - \frac{1}{5} \pi x^5 - \frac{2}{3} \pi x^3 + \pi x \right]_{-.7696426}^{.7696426}$

$= 1.7431 + 1.7431$

$= 3.4862$

Evaluate $\int_{0}^{\sqrt{3}} \frac{1}{\sqrt{9-x^2}} dx$.

(The **limit of an Improper Integral**. "If the limit exists, thee integral converges and the value of the integral is the limit. If the limit doesn't exist, the integral diverges and the integral cannot be evaluated there."

Solution: If any x value in [a, b] makes $f(x)$ undefined, then $\int_a^b f(x) dx$ is an improper integral. That's the situation here, since the integrand is undefined at the upper limit, $x = \sqrt{3}$. So, we rewrite the integral with the limit as $b$ approaches $\sqrt{3}$ from the left. Then we transform the integrand into sine inverse form by factoring out a 9 inside the radical and using a $u$-substitution, remembering to change our limits we do:

$\int_0^{\sqrt{3}} \frac{1}{\sqrt{9-x^2}} dx = \lim_{b \to \sqrt{3}^-} \int_0^b \frac{1}{\sqrt{9-x^2}} dx = \lim_{b \to \sqrt{3}^-} \int_0^b \frac{1}{3\sqrt{1-(\frac{1}{3}x)^2}} dx$

$\begin{cases} u = \frac{1}{3} x \\ du = \frac{1}{3} dx \end{cases}$

$$= \lim_{b \to \sqrt{3}^-} \int_0^{b/3} \frac{(3\,du)}{3\sqrt{1-u^2}} = \lim_{b \to \sqrt{3}^-} [\sin^{-1} u]_0^{b/3} = \lim_{b \to \sqrt{3}^-} [\sin^{-1}(\tfrac{1}{3}b) - \sin^{-1}(0)]$$

$$= \sin^{-1} \tfrac{\sqrt{3}}{3}.$$

**Partial fracs w/ diffeq**

Find the solution of the differential equation $\dfrac{dy}{dx} = \dfrac{y^3 + 3y^2 + y + 3}{4y+9}\left(\dfrac{e^x}{\sqrt{1-e^{2x}}}\right)$, with the initial condition: $y(0) = -1$.

(Use various kind of technique of integration to solve an initial value problem. The technique use here includes partial fraction, u-substitution and trig-substitution).

Solution: $\dfrac{dy}{dx} = \dfrac{y^3 + 3y^2 + y + 3}{4y+9}\left(\dfrac{e^x}{\sqrt{1-e^{2x}}}\right)$. Separating the variables and integrating,

$$\dfrac{4y+9}{y^3 + 3y^2 + y + 3}\,dy = \left(\dfrac{e^x}{\sqrt{1-e^{2x}}}\right)dx \Rightarrow \int \dfrac{4y+9}{y^3 + 3y^2 + y + 3}\,dy = \int \left(\dfrac{e^x}{\sqrt{1-e^{2x}}}\right)dx \quad (*)$$

To evaluate the left side of $\int \dfrac{4y+9}{y^3 + 3y^2 + y + 3}\,dy$, we use partial fractions:

$$\dfrac{4y+9}{(y+3)(y^2+1)} = \dfrac{A}{(y+3)} + \dfrac{By+C}{(y^2+1)}$$

$$4y+9 = A(y^2+1) + (By+C)(y+3)$$

$$4x+9 = Ay^2 + A + By^2 + 3By + Cy + 3C$$

$$4x+9 = (A+B)y^2 + (3B+C)y + (A+3C)$$

$$(A+B) = 0$$
$$(3B+C) = 4$$
$$(A+3C) = 9$$

$$A = \dfrac{-3}{10},\ B = \dfrac{3}{10},\ C = \dfrac{31}{10}$$

So $\int \dfrac{4y+9}{y^3 + 3y^2 + y + 3}\,dy$

$$= \int \frac{-3/10}{y+3} dy + \int \frac{\frac{3}{10}y + \frac{31}{10}}{y^2+1} dy$$

$$= \frac{-3}{10}\int \frac{1}{y+3} dy + \frac{31}{10}\int \frac{1}{y^2+1} dy + \frac{3}{10}\int \frac{y}{y^2+1} dy$$

$$u = y^2 + 1$$
$$du = 2y\,dy$$

$$= \frac{-3}{10}\ln|y+3| + \frac{31}{10}\tan^{-1}y + \frac{3}{20}\ln|y^2+1|\,dy$$

Next, evaluate the right hand side of (*),.

$$\int \left(\frac{e^x}{\sqrt{1-e^{2x}}}\right) dx$$

$$\begin{cases} u = e^x \\ du = e^x dx \end{cases}$$

$$= \int \frac{du}{\sqrt{1-u^2}}$$

$$= \sin^{-1} u + C$$

$$= \sin^{-1}(e^x) + C$$

Thus $\frac{-3}{10}\ln|y+3| + \frac{31}{10}\tan^{-1}y + \frac{3}{20}\ln|y^2+1| = \sin^{-1}(e^x) + C$.

Use the initial condition y(0)=-1,

$$\frac{-3}{10}\ln|2| + \frac{31}{10}\tan^{-1}(-1) + \frac{3}{20}\ln|2| = \sin^{-1}(e^0) + C$$

$$C \approx -1.6162$$

$$\frac{-3}{10}\ln|y+3| + \frac{31}{10}\tan^{-1}y + \frac{3}{20}\ln|y^2+1| = \sin^{-1}(e^x) - 1.6162$$

Using **integration by parts**, find $\int x\,\sec(x^2)\,dx$.

Solution: The formula that must be known in order to apply integration by parts is as follows:

$$\int u\,dv = uv - \int v\,du$$

$$\int x\,\sec^2(x)\,dx$$

$$\begin{array}{cc} \uparrow & \uparrow \\ u & \partial v \end{array}$$

$u = x \qquad dv = \sec^2(x)\, dx$
$du = dx \qquad v = \tan(x)$

✓ $\qquad x\tan(x) - \int \tan(x)\, dx \qquad$ *We must rewrite •tan (x) ∂x

$\int \tan(x)\, dx = \int \dfrac{\sin(x)}{\cos(x)}\, dx$

$z = \cos(x)$
$dz = -\sin(x)\, dx \qquad$ *Perform a simple substitution

$-\int \left(\dfrac{1}{z}\right) dz$

$-\ln(z) + C$

$\ln(\cos(x)) + C \qquad$ *Take answer and replace •tan (x) ∂x w/ it

⇨ $\qquad x\tan(x) + \ln(\cos(x)) + C \qquad$ *Don't forget minus times minus is positive

With your knowledge of **partial fractions**, solve this integral: $\int \dfrac{15x^4 - 2x^2 + 6}{x^4 + 3x^3 - 7x^2 - 21x}\, dx$

Solution: The first step to solving this problem would be to rewrite it.

$\int \dfrac{15x^4 - 2x^2 + 6}{x^4 + 3x^3 - 7x^2 - 21x}\, dx = \int \dfrac{15x^4 - 2x^2 + 6}{(x^2 + 3x)(x^2 - 7)}\, dx \qquad$ *Factor the numerator

$= \int \dfrac{15x^4 - 2x^2 + 6}{x(x + 3)(x^2 - 7)}\, dx \qquad$ *Factor out an x

$\dfrac{15x^4 - 2x^2 + 6}{x(x + 3)(x^2 - 7)} = \dfrac{A}{x} + \dfrac{B}{(x + 3)} + \dfrac{Cx + D}{(x^2 - 7)}$

$15x^4 - 2x^2 + 6 = A(x + 3)(x^2 - 7) + Bx(x^2 - 7) + (Cx + D)(x)(x + 3)$

$$\text{if } x = 0: \quad \begin{aligned} 6 &= A(3)(-7) \\ -\frac{2}{7} &= A \end{aligned} \qquad \text{if } x = -3: \quad \begin{aligned} 1215 - 18 + 6 &= B(-3)(2) \\ -200.5 &= B \end{aligned}$$

$$15x^4 - 2x^2 + 6 = A(x^3 + 3x^2 - 7x - 21) + B(x^3 - 7x) + C(x^3 + 3x^2) + D(x^2 + 3x)$$

$$0x^3 = Ax^3 + Bx^3 + Cx^3 \qquad\qquad -2x^2 = 3Ax^2 + 3Cx^2 + Dx^2$$

$$0 = -\frac{2}{7} - \frac{401}{2} + C \qquad\qquad -2 = 3\left(-\frac{2}{7}\right) + 3\left(\frac{2811}{14}\right) + D$$

$$\frac{2811}{14} = C \qquad\qquad -\frac{1207}{2} = D$$

So, rewriting the integral

$$\int \left[ \frac{-2/7}{x} + \frac{-401/2}{(x+3)} + \frac{(2811/14)x + (-1207/2)}{(x^2 - 7)} \right] dx$$

$$\Rightarrow \int \left[ \frac{-2/7}{x} + \frac{-401/2}{(x+3)} + \frac{(2811/14)x}{(x^2 - 7)} + \frac{(-1207/2)}{(x^2 - 7)} \right] dx$$

$$\Rightarrow -\frac{2}{7}\ln|x| - \frac{401}{2}\ln|x+3| + \frac{2811}{28}\ln|x^2 - 7| - \frac{1207}{2} \int \frac{1}{(x^2 - 7)} dx$$

Integrate: $\int \sec^4(x) \tan^4(x) dx$

Solution: $\int \sec^4(x) \tan^4(x) dx = \int \sec^2(x) \tan^4(x) \sec^2(x) dx$

$\Rightarrow \int (\tan^2(x) + 1) \tan^4(x) \sec^2(x) dx$

$\quad u = \tan(x) \quad du = \sec^2(x)$

$\Rightarrow \int (u^6 + u^4) du$

$\Rightarrow \frac{1}{7}u^7 + \frac{1}{5}u^5 + C = \frac{1}{7}\tan^7(x) + \frac{1}{5}\tan^5(x) + C$

Integrate: $\boxed{\int \frac{1}{(x-4)(x-2)(x-3)} dx}$

Solution:

$$= \frac{A}{(x-4)} + \frac{B}{(x-2)} + \frac{C}{(x-3)}$$

$\Rightarrow 1 = A(x-2)(x-3) + B(x-4)(x-3) + C(x-2)(x-4)$

$\Rightarrow x = 4; A = 1/2 \qquad x = 2; B = 1/2 \qquad x = 3; C = -1$

$$= \frac{1}{2}\int \frac{1}{x-4}dx + \frac{1}{2}\int \frac{1}{x-2}dx - \int \frac{1}{x-3}dx$$

$$= \frac{1}{2}\ln|x-4| + \frac{1}{2}\ln|x-2| - \ln|x-3| + C$$

$\int x^3 \cos(x)dx$ **NO TABULAR METHOD!**

Solution: $u = x^3 \quad dv = \cos(x)dx$
$du = 3x^2 \quad v = \sin(x)$

$$= x^3 \sin(x) - 3\int x^2 \sin(x)dx$$

$u = x^2 \quad dv = \sin(x)dx$
$du = 2x \quad v = -\cos(x)$

$$= x^3 \sin(x) - 3x^2 \cos(x) + 6\int x\cos(x)dx$$

$u = x \quad dv = \cos(x)dx$
$du = dx \quad v = \sin(x)$

$$= x^3 \sin(x) - 3x^2 \cos(x) + 6x\sin(x) - 6\int \sin(x)dx$$

$$= x^3 \sin(x) - 3x^2 \cos(x) + 6x\sin(x) + 6\cos(x) + C$$

This problem uses the method of **integration by parts,** and just as a warning do not give up in the middle of the problem, you will have to use the method twice in order to get to an answer.

$\int x^2 \cos x\, dx$

Solution: Remember that integration by parts is $\int u\,dv = uv - \int v\,du$, so first determine what $u$ and $dv$ should be.
Let $u = x^2$ and $dv = \cos x\,dx$
So $du = 2x\,dx$ and $v = \sin x$

$$\int x^2 \cos x\,dx = x^2 \sin x - \int \sin x\, 2x\,dx$$

Can you spot the problem yet?...okay I'll tell you, we need to use integration by parts again for the antiderivative of sinx2x.
This time let $u = 2x$ and $dv = \sin x\,dx$
Which means $du = 2\,dx$ and $v = -\cos x$

$$\int 2x\sin x\,dx = -2x\cos x - \int -\cos x\, 2\,dx = -2x\cos x + 2\int \cos x\,dx = -2x\cos x + 2\sin x$$

Put both of the answers together to get the final answer of
$$\int x^2 \cos x\,dx = x^2 \sin x + 2x\cos x - 2\sin x$$

**Basic hyperbolic inverse functions & trig substitution** Find $\int \frac{dx}{\sqrt{9+9x^2}}$ using trig substitution and then using hyperbolic functions.

Solution: **Add trig sub here**

Method #2: $\frac{d}{dx}\sinh^{-1} x = \frac{1}{\sqrt{1+x^2}}$  In this particular problem you can factor out the 9 to get it into the general form.

$$\int \frac{dx}{\sqrt{9+9x^2}} = \frac{1}{3}\int \frac{dx}{\sqrt{1+x^2}} = \frac{1}{3}\sinh^{-1} x + c$$

**integration by parts using the tabular method**: Find $\int x^3 \cos x\, dx$.

Solution:

|   | u | $\frac{dv}{dx}$ |
|---|---|---|
| (+) | $x^3$ | cos x |
| (−) | $3x^2$ | sin x |
| (+) | $6x$ | −cos x |
| (−) | 6 | −sin x |
| 0 |   | cos x |

$$\int x^3 \cos x\, dx = x^3 \sin x + 3x^2 \cos x - 6x\sin x - 6\cos x + C$$

Check:
$$\frac{d}{dx}(x^3 \sin x + 3x^2 \cos x - 6x \sin x - 6\cos x + C)$$
$$= (3x^2 \sin x + x^3 \cos x) + (6x\cos x - 3x^2 \sin x) - (6\sin x + 6x\cos x) - (-6\sin x)$$
$$= x^3 \cos x$$

**Integration by parts. get integral in terms of itself** Find $\int e^{3x} \cos 7x\, dx$.

Solution: Because the problem involves two distinct parts that can easily be integrated separately, integration by parts will be used. The tabular method could not be used in this case, since the derivatives of both $e^{3x}$ and cos7x never go to zero. In this example, *u* was chosen to be $e^{3x}$, but it could have also been cos7x.

$\begin{bmatrix} u = e^{3x} & du = 3e^{3x}dx \\ dv = \cos 7x & v = \frac{1}{7}\sin 7x \end{bmatrix}$ $\begin{array}{l} \int u\,dv = uv - \int v\,du \\ \Rightarrow \int e^{3x} \cos 7x\, dx = e^{3x}(\tfrac{1}{7}\sin 7x) - \int (\tfrac{1}{7}\sin 7x)(3e^{3x})dx \end{array}$ Because the integration $\int(\tfrac{1}{7}\sin 7x)(3e^{3x})dx$ still has two distinct parts, integration by parts should be used again. To avoid undoing the work just done, *u* should be *du* from the first integration, and *dv* should be *v*.

$\begin{bmatrix} u = 3e^{3x} & du = 9e^{3x}dx \\ dv = \frac{1}{7}\sin 7x dx & v = -\frac{1}{49}\cos 7x \end{bmatrix} \Rightarrow \int (\frac{1}{7}\sin 7x)(3e^{3x})dx = 3e^{3x}$
$(-\frac{1}{49}\cos 7x) + \frac{9}{49}\int \cos 7x e^{3x}dx$

↑ I=$\int \cos 7x e^{3x}dx$. Now, notice the integration that's left is the same as the original, but multiplied by 9/49, so call it "I" and solve for it.

$I = e^{3x}(\frac{1}{7}\sin 7x) - 3e^{3x}(-\frac{1}{49}\cos 7x) - \frac{9}{49}I$

$\frac{58}{49}I = \frac{1}{7}e^{3x}\sin 7x + \frac{3}{49}e^{3x}\cos 7x + C$

$I = \frac{49}{58}(\frac{1}{7}e^{3x}\sin 7x + \frac{3}{49}e^{3x}\cos 7x) + C$

$I = \frac{7}{58}e^{3x}\sin 7x + \frac{3}{58}\cos 7x + C$

$\Rightarrow \int e^{3x}\cos 7x dx = \frac{7}{58}e^{3x}\sin 7x + \frac{3}{58}\cos 7x + C$


): **partial fractions. matrices** Find $\int \frac{4x^3 + 7x^2 + 6x + 12}{(x^2 - 3)(x+1)^2}dx$.

Solution: $\frac{4x^3 + 7x^2 + 6x + 12}{(x^2 - 3)(x+1)^2 (x)} = \frac{Ax + B}{(x^2 - 3)} + \frac{C}{(x+1)} + \frac{D}{(x+1)^2} + \frac{E}{x}$. Multiply both sides by $(x^2 - 3)(x+1)^2$ and reduce to obtain:

$4x^3 + 7x^2 + 6x + 12 = (Ax + B)(x)(x+1)^2 + C(x)(x+1)(x^2 - 3) + D(x)(x^2 - 3) + E(x^2 - 3)(x+1)^2$

. Plug in various values for $x$ that are useful in negating terms:

→ $x = -1$: $-4 + 7 - 6 + 12 = 2D$     $D = 9/2$

→ $x = 0$: $12 = -3E$     $E = -4$

→ $x = \sqrt{3}$: $(\sqrt{3}A + B)(\sqrt{3})(\sqrt{3} + 1)^2 = 12\sqrt{3} + 21 + 6\sqrt{3} + 12$

   $3A + \sqrt{3}B = \frac{18\sqrt{3} + 33}{(\sqrt{3}-1)^2}$

→ $x = 1$: $4(A + B) + 8C - 2D - 8E = 4 + 7 + 6 + 12$

   $4A + 4B + 8C - 2(9/2) - 8(-4) = 29$

   $A + B + 2C = 3/2$

→ $x = 2$: $(2A + B)(9)(2) + C(2)(3)(7) + D(2)(1) + E(1)(9) = 4(2)^3 + 7(4) + 12 + 12$

   $36A + 18B + 42C + 2(9/2) + 9(-4) = 84$

   $36A + 18B + 42C = 111$

Solving this set of equations could be potentially nerve-racking, therefore it is very useful to use your graphing calculator and solve them with a matrix:

Equations: $\begin{array}{l} A + B + 2C = 3/2 \\ 36A + 18B + 42C = 111 \\ 3A + \sqrt{3}B \approx 8.6 \end{array}$  Matrix form: $\begin{bmatrix} 1 & 1 & 2 & 3/2 \\ 36 & 18 & 42 & 111 \\ 3 & \sqrt{3} & 0 & \approx 8.6 \end{bmatrix}$

To use the calculator, put the matrix in and then use "rref" to solve for it in the form of an identity matrix, the solutions to *A*, *B*, and *C* consecutively will be on the right.

$$A = 4.67377 \approx 4.7$$
$$B = -3.1311 \approx -3.1$$
$$C = -0.02132 \approx -0.02$$
$$D = 3/2$$
$$E = -4$$

$$\Rightarrow \int \frac{Ax+B}{x^2-3}dx + (C)\int \frac{1}{1+x}dx + (D)\int \frac{1}{(x+1)^2}dx + (E)\int \frac{1}{x}dx$$

$$\Rightarrow (A)\int \frac{x}{x^2-3}dx + (B)\int \frac{1}{x^2-3}dx + (C)\int \frac{1}{1+x}dx + (D)\int \frac{1}{(x+1)^2}dx + (E)\int \frac{1}{x}dx$$

$$\uparrow [u = x^2-3,\ du = 2x] \quad \uparrow [\tanh^{-1}(\tfrac{x}{a}) = \tfrac{1}{a}\int \tfrac{1}{a^2-x^2}dx] \quad \uparrow [v = x+1,\ dv = dx]$$

$$\Rightarrow (\tfrac{A}{2})\int \frac{1}{u}du - (B)\int \frac{1}{(\sqrt{3})^2-x^2}dx + (C)\int \frac{1}{v}dv + (D)\int \frac{1}{v^2}dv + (E)\int \frac{1}{x}dx$$

$$\Rightarrow \tfrac{4.7}{2}\ln(x^2-3) - \tfrac{(-3.1)}{\sqrt{3}}\tanh^{-1}\left(\tfrac{x}{\sqrt{3}}\right) + (-0.02)\ln(x+1) - 3(3/2)x^{-3} + (-4)\ln(x) + c$$

$$\Rightarrow 2.35 \cdot \ln(x^2-3) + 1.79 \cdot \tanh^{-1}\left(\frac{x}{\sqrt{3}}\right) - 0.02 \cdot \ln(x+1) - \frac{27}{2x^3} - 4 \cdot \ln(x) + c$$

**L'Hopital's Theorem. using natural log** Find $\lim\limits_{x \to \infty} x^{2/x^2}$.

Solution: The limit is currently in the indeterminate form of $\infty^0$, which is a form that L'Hopital's theorem can't be used on. However, by calling the limit "f(x)" and taking it's natural log, it could be transformed into a form that L'Hopital's theorem can be used on. This can only be done because the natural log function is a one-to-one function.

$$f(x) = \lim_{x \to \infty} x^{\frac{2}{x^2}}$$

$$\ln f(x) = \ln(\lim_{x \to \infty} x^{\frac{2}{x^2}}) \qquad * f(\lim_{x \to c} g(x)) = \lim_{x \to c} f(g(x))$$

$$= \lim_{x \to \infty} \ln(x^{\frac{2}{x^2}})$$

$$= \lim_{x \to \infty} \tfrac{2}{x^2} \ln x = 2 \lim_{x \to \infty} \tfrac{\ln x}{x^2}$$

Now that the limit is in the form $\tfrac{\infty}{\infty}$, L'Hopital's theorem can be used:

$$\lim_{x \to \infty} \tfrac{1/x}{2x} = 2 \lim_{x \to \infty} \tfrac{1}{2x^2} = 2(0) = 0$$

So, ln f(x)=0 and **f(x)=1**.

**integration by parts** Evaluate $\int xe^{9x}dx$.

Solution: To find the solution to this problem integration by parts and the formula, $\int udv = uv - \int vdv$, needs to be used. In order to do so we will first substitute a value for *u*, in this case $x$, then the value for $dv$ will be the rest of the initial integral, $e^{9x}$. Then, the values of $du$ and $v$ can then be derived and the equation solved using the before mentioned formula.

$$\int u\,dv = uv - \int v\,dv$$

Let $u = x$ & $dv = e^{9x}$

Then, $du = dx$ & $v = \dfrac{1}{9}e^{9x}$

$$\int xe^{9x}\,dx = x\left(\dfrac{1}{9}e^{9x}\right) - \int \dfrac{1}{9}e^{9x}$$

$$\int xe^{9x}\,dx = \dfrac{x}{9}e^{9x} - \dfrac{1}{81}e^{9x} + C$$

The most important thing to be able to do in these problems is to designate good values for $u$ and $dv$. A nice rule of thumb when dealing with problems involving $e$ multiplied by another variable, such as we have here, is to make the other variable $u$, and designate the portion of the integral with e in it as $dv$.

Using **L'Hopital's Rule**, find $\lim\limits_{x \to 0} \dfrac{\sin x}{x}$.

Solution: This is a $0/0$ indeterminant form, so L'Hopital's rule applies:

$$\lim_{x \to 0} \dfrac{\sin x}{x} = \lim_{x \to 0} \dfrac{\dfrac{d}{dx}(\sin x)}{\dfrac{d}{dx}(x)} = \lim_{x \to 0} \dfrac{\cos x}{1} = 1$$

**Derive the integration by parts formula.**

Let $u$ and $v$ each be a function of $x$. Using the product rule, $\dfrac{d}{dx}(uv) = u\dfrac{dv}{dx} + v\dfrac{du}{dx}$

$\Rightarrow u\dfrac{dv}{dx} = \dfrac{d}{dx}(uv) - v\dfrac{du}{dx}$. Multiplying through by $dx$, we have $u\,dv = d(uv) - v\,du$. Finally, integrating both sides of the equation yields $\int u\,dv = \int d(uv) - \int v\,du \Rightarrow \int u\,dv = uv - \int v\,du$.

|   | $u$ | $dv/dx$ |
|---|---|---|
| + | $x^3$ | $\cos 2x$ |
| - | $3x^2$ | $\dfrac{1}{2}\sin 2x$ |
| + | $6x$ | $-\dfrac{1}{4}\cos 2x$ |

**integration by parts and the tabular method.** Find $\int x^3 \cos 2x\, dx$.

Solution: Since the integrand is comprised of a polynomial and a trig function, we let $u$ = the polynomial and $dv$ = the trig function times $dx$. Since repeated differentiation of $u$ will eventually yield zero, this problem can be handled quickly with the tabular method. (Without it we would have to use integration by parts three times.)

| | | |
|---|---|---|
| $x^3$ | | $\cos 2x$ |
| $3x^2$ | − | $\frac{1}{2}\sin 2x$ |
| $6x$ | + | $-\frac{1}{4}\cos 2x$ |
| $6$ | − | $-\frac{1}{8}\sin 2x$ |
| $0$ | | $\frac{1}{16}\cos 2x$ |

Thus, $\int x^3 \cos 2x\, dx$

$= x^3 \cdot \frac{1}{2}\sin 2x - 3x^2 \cdot \frac{-1}{4}\cos 2x + 6x \cdot \frac{-1}{8}\sin 2x - 6 \cdot \frac{1}{16}\cos 2x + C$

$= \frac{x^3 \sin 2x}{2} + \frac{3x^2 \cos 2x}{4} - \frac{6x \sin 2x}{8} - \frac{3 \cos 2x}{8} + C$.

**trig integral** Find $\int 5\sin^5 x\, dx$.

Solution: If we had the same integral with a factor of $\cos x$, we would have a simple $u$-substitution problem by letting
$u = \sin x$, yielding $du = \cos x\, dx$. However, with this factor of $\cos x$, it is a more difficult problem. Since we have an odd power of $\sin x$, we break one off to become part of $du$. Then we let $u = \cos x$, so $du = -\sin x\, dx$:

$\int 5\sin^5 x\, dx = 5\int (1 - \cos^2 x)^2 \sin x\, dx = -5\int (1 - u^2)^2 du = -5\int (1 - 2u^2 + u^4)\, du$

$= -5\int du + 10\int u^2 du - 5\int u^4 du = -5u + \frac{10}{3}u^3 - u^5 + C = -5\cos x + \frac{10}{3}\cos^3 x - \cos^5 x + C$

Evaluate the limit: $\lim_{x \to 0} \frac{e^{12x} - 1}{2x}$.

Solution: Both the numerator and denominator approach zero as $x$ does, so we have a $0/0$ indeterminate situation for which L'Hospital's rule applies. So, $\lim_{x \to 0} \frac{e^{12x} - 1}{2x} = \lim_{x \to 0} \frac{\frac{d}{dx}(e^{12x} - 1)}{\frac{d}{dx}(2x)} =$

$\lim_{x \to 0} \frac{12e^{12x}}{2} = 6e^0 = 6$.

**partial fractions** Find $\int \frac{x^2 + 22x + 121}{x^6 + 24x^5 + 163x^4 + 192x^3 + 619x^2 + 742x + 363}\, dx$.

Solution: We begin by factoring the numerator and denominator. For the latter we note that $363 = 3(11)^2$. By the rational root theorem, if there are any rational roots to the denominator, they must be among $\{\pm 1, \pm 3, \pm 11\}$, since the leading coefficient is 1. (Note that there could be other irrational or nonreal roots.) By Descartes' rule of signs we can exclude any positive roots and conclude that there are 0, 2, 4, or 6 negative roots. Using synthetic division we find that

-1 and -11 are double roots. Thus the integrand factors into $\frac{(x+11)^2}{(x+11)^2 (x+1)^2 (x^2+3)} =$

$\dfrac{1}{(x+1)^2(x^2+3)}$, $x \neq -1$ 1. To use partial fractions we rewrite the integrand:

$\dfrac{1}{(x+1)^2(x^2+3)} = \dfrac{A}{x+1} + \dfrac{B}{(x+1)^2} + \dfrac{Cx+D}{x^2+3}$. Since the denominator has a linear factor squared, we must list it twice, once to the first power, and once to the second. Each linear factor has an unknown constant above it. The second degree factor requires an known linear function above it. So, we have four unknown for which we need to solve. Multiplying through by $(x+1)^2(x^2+3)$, we obtain $1 = A(x+1)(x^2+3) + B(x^2+3) + (Cx+D)(x+1)^2$. When $x = -1$, we have $1 = 4B$, or $B = \frac{1}{4}$. Substituting, $1 = A(x+1)(x^2+3) + \dfrac{1}{4}(x^2+3) + (Cx+D)(x+1)^2$. To solve for the remaining three unknowns, we multiply through by 4, expand, and combine like terms:

$4 = 4A(x+1)(x^2+3) + (x^2+3) + 4(Cx+D)(x+1)^2$
$= (12A + 12Ax + 4Ax^2 + 4Ax^3) + (x^2+3) + (4D + 4Cx + 8Dx + 8Cx^2 + 4Dx^2 + 4Cx^3)$
$= (3 + 12A + 4D) + (12A + 4C + 8D)x + (1 + 4A + 8C + 4D)x^2 + (4A + 4C)x^3$. For two polynomials to be equal, all coefficients of all powers of $x$ must be equal. Since this entire polynomial is equal to 4, its constant term must equal 4, while all other coefficients must equal zero. This results in the following system of linear equations:

$$3 + 12A + 4D = 4, \quad 12A + 4C + 8D = 0, \quad 1 + 4A + 8C + 4D = 0, \quad 4A + 4C = 0$$

We have four independent equations relating three unknowns, $A$, $C$, and $D$ (since we've already found $B$), so we can use any three of the equations to find a solution to the system. Creating an augmented matrix with the first, second, and fourth equations, and using row operations to simplify it, we have:

$\begin{bmatrix} 12 & 0 & 4 & 1 \\ 12 & 4 & 8 & 0 \\ 4 & 4 & 0 & 0 \end{bmatrix} \sim \begin{bmatrix} 12 & 0 & 4 & 1 \\ 3 & 1 & 2 & 0 \\ 1 & 1 & 0 & 0 \end{bmatrix} \sim \begin{bmatrix} 1 & 1 & 0 & 0 \\ 3 & 1 & 2 & 0 \\ 12 & 0 & 4 & 1 \end{bmatrix} \sim \begin{bmatrix} 1 & 1 & 0 & 0 \\ 0 & -2 & 2 & 0 \\ 0 & -12 & 4 & 1 \end{bmatrix} \sim \begin{bmatrix} 1 & 1 & 0 & 0 \\ 0 & 1 & -1 & 0 \\ 0 & -12 & 4 & 1 \end{bmatrix}$

$\sim \begin{bmatrix} 1 & 0 & 1 & 0 \\ 0 & 1 & -1 & 0 \\ 0 & 0 & -8 & 1 \end{bmatrix} \sim \begin{bmatrix} 1 & 0 & 1 & 0 \\ 0 & 1 & -1 & 0 \\ 0 & 0 & 1 & -1/8 \end{bmatrix} \sim \begin{bmatrix} 1 & 0 & 1 & 0 \\ 0 & 1 & -1 & 0 \\ 0 & 0 & 1 & -1/8 \end{bmatrix} \sim \begin{bmatrix} 1 & 0 & 0 & 1/8 \\ 0 & 1 & 0 & -1/8 \\ 0 & 0 & 1 & -1/8 \end{bmatrix}$

The final row equivalent matrix is in reduced row echelon form, and clearly shows that $A = 1/8$, $C = -1/8$, and $D = -1/8$. We had already determined that $B = 1/4$. Substituting for these values we obtain

$\dfrac{1}{(x+1)^2(x^2+3)} = \dfrac{1/8}{x+1} + \dfrac{1/4}{(x+1)^2} + \dfrac{-x/8 - 1/8}{x^2+3}$. Integrating term by term yields

$\int \dfrac{1}{(x+1)^2(x^2+3)} dx = \dfrac{1}{8}\int \dfrac{1}{x+1} dx + \dfrac{1}{4}\int \dfrac{1}{(x+1)^2} dx - \dfrac{1}{8}\int \dfrac{x+1}{x^2+3} dx$. The first integral on the right is a simple natural log. For the second we let $u = x+1$, implying $du = dx$. We split the third integral into two. Thus, $\int \dfrac{1}{(x+1)^2(x^2+3)} dx = \dfrac{1}{8}\ln|x+1| + \dfrac{1}{4}\int u^{-2} du - \dfrac{1}{8}\int \dfrac{x}{x^2+3} dx - \dfrac{1}{8}\int \dfrac{1}{x^2+3} dx$.

For the second integral we have $\dfrac{1}{4}\int u^{-2} du = -\dfrac{1}{4}u^{-1} = \dfrac{-1}{4(x+1)}$. For the third integral we let $v = x^2 + 3$, implying $dv = 2x\, dx$. So, $-\dfrac{1}{8}\int \dfrac{x}{x^2+3} dx = -\dfrac{1}{16}\int \dfrac{2x}{x^2+3} dx = -\dfrac{1}{16}\int \dfrac{dv}{v} = -\dfrac{1}{16}\ln|v|$

$= -\frac{1}{16}\ln(x^2 + 3)$. (Absolute values are unneeded here since $x^2 + 3$ is always positive.) Finally, the last integral is a tangent inverse: $\int \frac{dx}{a^2 + x^2} = \frac{1}{a}\tan^{-1}\left(\frac{x}{a}\right) \Rightarrow -\frac{1}{8}\int \frac{dx}{x^2 + 3} =$

$-\frac{1}{8\sqrt{3}}\tan^{-1}\left(\frac{x}{\sqrt{3}}\right)$. Putting it all together: $\int \frac{1}{(x+1)^2(x^2+3)} dx = \frac{1}{8}\ln|x+1| - \frac{1}{4(x+1)}$

$-\frac{1}{16}\ln(x^2+3) - \frac{1}{8\sqrt{3}}\tan^{-1}\left(\frac{x}{\sqrt{3}}\right) + C$. Since $\frac{1}{16}\ln(x^2+3) = \frac{1}{8}\cdot\frac{1}{2}\ln(x^2+3) = \frac{1}{8}\ln\sqrt{x^2+3}$,

we can combine the two logarithms to rewrite our answer as

$\int \frac{1}{(x+1)^2(x^2+3)} dx = \frac{1}{8}\ln\frac{|x+1|}{\sqrt{x^2+3}} - \frac{1}{8\sqrt{3}}\tan^{-1}\left(\frac{x}{\sqrt{3}}\right) + C$.

What is the rational root theorem?
Explain how Descartes' rule of signs was used to exclude positive roots.
How does the fundamental theorem of algebra apply to this factoring the denominator.
Do the synthetic division not shown in the problem above in order to factor the denominator.

Apart Function


**integration by parts, trig identities** Find
$\frac{1}{\sqrt{2}}\int e^x \csc x \sqrt{(1 - \sin^2 x - \cos 2x)(1 + \cos^2 x - \sin^2 x)}\, dx$

Solution: Let's begin by simplifying the integrand. Using the double angle formula for cosine,
$1 - \sin^2 x - \cos 2x$
$= 1 - \sin^2 x - (\cos^2 x - \sin^2 x) = 1 - \cos^2 x = \sin^2 x$. For the second factor of the integrand we use the fact that $1 - \sin^2 x = \cos^2 x$ to obtain $2\cos^2 x$. Our integral now becomes

$\frac{1}{\sqrt{2}}\int e^x \csc x \sqrt{(\sin^2 x)(2\cos^2 x)}\, dx$

$= \frac{1}{\sqrt{2}}\int e^x \csc x \cdot \sqrt{2}\sin x \cos x\, dx = \int e^x \cos x\, dx$. Now it's time for integration by parts:

$\begin{cases} u = e^x & dv = \cos x \\ du = e^x dx & v = \sin x \end{cases} \Rightarrow \int e^x \cos x\, dx = e^x \sin x - \int e^x \sin x\, dx$. To find the integral on the

right we use parts a second time: $\begin{cases} u = e^x & dv = \sin x \\ du = e^x dx & v = -\cos x \end{cases} \Rightarrow$

$\int e^x \cos x\, dx = e^x \sin x - \left(-e^x \cos x + \int e^x \cos x\, dx\right)$

$\int e^x \cos x\, dx = e^x \sin x + e^x \cos x - \int e^x \cos x\, dx$. We now have the integral we're seeking in terms of itself, so now it's just a matter of algebraically solving for the integral:
$2\int e^x \cos x\, dx = e^x \sin x + e^x \cos x$

$\Rightarrow \int e^x \cos x\, dx = \frac{e^x}{2}(\sin x + \cos x) + C$.

**integration by parts** Find $\int (1-x)e^x\, dx$.

Solution: Using integration by parts, $\int udv = uv - \int vdu$

so, u= 1-x      dv= $e^x dx$
du= -dx      $\int dv = \int e^x (dx)$      $v = e^x$

Plug it in

$\int (1-x)e^x dx = (1-x)e^x - \int e^x (-dx)$

$= (1-x)e^x + \int e^x dx$

$= e^x - xe^x + e^x + c$

$= 2e^x - xe^x + c$

$= e^x (2-x) + c$

**integration by parts** Find the antiderivative of $e^x \cos x$.

Solution: Step 1: Use u-substitution

$\int e^x \cos x dx$    $u = \cos x$    $du = -\sin x dx$    $v = e^x$    $dv = e^x dx$

Use $\int udv = uv - \int vdu$

$\int e^x \cos x dx = e^x \cos x + \int e^x \sin x dx$

Step 2: Do substitution again with new integral:    $u = \sin x$    du=cosxdx    $v = e^x$    $dv = e^x dx$

$e^x \cos x + e^x \sin x - \int e^x \cos x dx$

Step3: Set original equation equal to $e^x \cos x + e^x \sin x - \int e^x \cos x dx$

$\int e^x \cos x dx = e^x \cos x + e^x \sin x - \int e^x \cos x dx$

Step 4: Solve for $\int e^x \cos x dx$ using algebra

$2 \int e^x \cos x dx = e^x \cos x + e^x \sin x + C$

$\int e^x \cos x dx = \dfrac{e^x \cos x + e^x \sin x}{2} + C$


**trig, partial fractions, special factoring, Mathematica** Find the antiderivative of $\sqrt{\tan x}$.

First have students expand $(u^2 + u\sqrt{a} + 1)(u^2 - u\sqrt{a} + 1)$ and show that if this expansion equal $u^4 + 1$, then $a = 2$.


Solution: *Step1 U-substitution; Step2 $\sec^2 x = \tan^2 x + 1$ & substitute; Step3 Plug in U for $\sqrt{\tan x}$; Step4: Solve for dx; Step5 Plug in $\dfrac{2u}{u^4 + 1}$ for dx; Step6; Factor $u^4 + 1$; Step7 Getting partial fractions, Multiplying $u^4 + 1$ by each side, Then foil out the rest, Put like terms together, Set coefficients equal, Solve for A,B,C and D; Step8 Substitute A,B,C and D into original equation; Step9 Split into two integrals; Step10 Perfect square polynomials; Step11;*

*Setting up for arctan integral* $\left(\int \frac{dx}{x^2+1} = \arctan x + c\right)$; *Step12 Substitution U and du; Step13 Break up into 4 integrals; Step14 Solving each of the 4 integrals; Step15 Combine ln's; Step16 Back substituting, Plug in expressions for V&W that are in terms of U, Square out* $(\sqrt{2}u-1)^2$ *and* $(\sqrt{2}u+1)^2$, *Simplify 1+1=2, Cancel the 2's, Plug in expression for u* $u = \sqrt{\tan x}$, *Simplify.*

$$u = \sqrt{\tan x} \qquad du = \frac{\sec^2 x}{2\sqrt{\tan x}}dx = \frac{\tan^2 x + 1}{2\sqrt{\tan x}}dx = \frac{u^4+1}{2u}dx \implies dx = \frac{2u}{u^4+1}du$$

Thus, $\int \sqrt{\tan x}\, dx = \int u\, dx = \int \frac{2u^2}{u^4+1}du$

Step6: $u^4 + 1 = (u^2 + u\sqrt{2} + 1)(u^2 - u\sqrt{2} + 1)$

Step 7: $\dfrac{2u^2}{u^4+1} = \dfrac{Au+B}{(u^2+u\sqrt{2}+1)} + \dfrac{Cu+D}{(u^2-u\sqrt{2}+1)}$

$2u^2 = (Au+B)(u^2 - u\sqrt{2} + 1) + (Cu+D)(u^2 + u\sqrt{2} + 1)$

$2u^2 = Au^3 - \sqrt{2}Au^2 + Au + Bu^2 - \sqrt{2}Bu + B + Cu^3 + \sqrt{2}Cu^2 + Cu + Du^2 + D\sqrt{2}u + D$

$2u^2 = (A+C)u^3 + (-\sqrt{2}A + B + \sqrt{2}C + D)u^2 + (A - \sqrt{2}B + C + \sqrt{2}D)u + B + D$

$A+C=0 \quad -\sqrt{2}A + B + \sqrt{2}C + D = 2 \quad A - \sqrt{2}B + C + \sqrt{2}D = 0 \quad B+D=0$

Solve for A, B, C and D : *=original equation
*A+C=0  *$A - \sqrt{2}B + C + \sqrt{2}D = 0$  $A + C - \sqrt{2}B + \sqrt{2}D = 0$  $-\sqrt{2}B + \sqrt{2}D = 0$
$B - D = 0$  B=D  done.
*B+D=0  B+B=0  2B=0  B=0  D=0  done.

*$-\sqrt{2}A + B + \sqrt{2}C + D = 2$  $-\sqrt{2}A + \sqrt{2}C = 2$  $-A+C= \sqrt{2}$

*A+C=0  $2C= \sqrt{2}$  $C = \dfrac{\sqrt{2}}{2}$  $A = \dfrac{-\sqrt{2}}{2}$

Step 8: $\dfrac{2u^2}{u^4+1} = \dfrac{-1}{\sqrt{2}}\left(\dfrac{u}{u^2+\sqrt{2}u+1}\right) + \dfrac{1}{\sqrt{2}}\left(\dfrac{u}{u^2-\sqrt{2}u+1}\right)$

Step 9: $\int \sqrt{\tan x}\,dx = \dfrac{-1}{\sqrt{2}}\int \dfrac{u}{u^2+\sqrt{2}u+1}du + \dfrac{1}{\sqrt{2}}\int \dfrac{u}{u^2-\sqrt{2}u+1}du$

$\int \sqrt{\tan x}\,dx = \dfrac{1}{\sqrt{2}}\int \dfrac{u}{u^2-\sqrt{2}u+1}du - \dfrac{1}{\sqrt{2}}\int \dfrac{u}{u^2+\sqrt{2}u+1}du$

Step 10: $\int \sqrt{\tan x}\,dx = \dfrac{1}{\sqrt{2}}\int \dfrac{u}{u^2 - \sqrt{2}u + \dfrac{1}{2} + \dfrac{1}{2}}(du) - \dfrac{1}{\sqrt{2}}\int \dfrac{u}{u^2 + \sqrt{2}u + \dfrac{1}{2} + \dfrac{1}{2}}$

$$\int \sqrt{\tan x}\,dx = \dfrac{1}{\sqrt{2}}\int \dfrac{u}{(u^2 - \dfrac{1}{\sqrt{2}})^2 + \dfrac{1}{2}}\,du - \dfrac{1}{\sqrt{2}}\int \dfrac{u}{(u^2 + \dfrac{1}{\sqrt{2}})^2 + \dfrac{1}{2}}\,du$$

Step 11: $\int \sqrt{\tan x}\,dx = \dfrac{1}{\sqrt{2}}\int \dfrac{2u}{2(u - \dfrac{1}{\sqrt{2}})^2 + 1}\,du - \dfrac{1}{\sqrt{2}}\int \dfrac{2u}{2(u + \dfrac{1}{\sqrt{2}})^2 + 1}\,du$

Make the equation look like $x^2 + 1$

$$\int \sqrt{\tan x}\,dx = \sqrt{2}\int \dfrac{2u}{\sqrt{2}(u - \dfrac{1}{\sqrt{2}})^2 + 1}\,du - \sqrt{2}\int \dfrac{2u}{\sqrt{2}(u + \dfrac{1}{\sqrt{2}})^2 + 1}\,du$$

Simplify: $\int \sqrt{\tan x}\,dx = \sqrt{2}\int \dfrac{u}{(\sqrt{2}u - 1)^2 + 1}\,du - \sqrt{2}\int \dfrac{u}{(\sqrt{2}u + 1)^2 + 1}\,du$

Step 12: $v + 1 = \sqrt{2}u,\ v = \sqrt{2}u - 1 \qquad \dfrac{v+1}{\sqrt{2}} = u$

Step 13: $\int \sqrt{\tan x}\,dx = \dfrac{1}{\sqrt{2}}\int \dfrac{v}{v^2+1}\,dv + \dfrac{1}{\sqrt{2}}\int \dfrac{dv}{v^2+1} - \dfrac{1}{\sqrt{2}}\int \dfrac{w}{w^2+1}\,dw + \dfrac{1}{\sqrt{2}}\int \dfrac{dw}{w^2+1}$

$\dfrac{1}{2\sqrt{2}}\ln(u^2+1) \qquad \downarrow$

$\dfrac{1}{\sqrt{2}}\arctan u$

Step 14:
$s = v^2 + 1$
$ds = 2v\,dv$

$\dfrac{1}{2\sqrt{2}}\int \dfrac{ds}{s} = \dfrac{1}{2\sqrt{2}}\ln s = \dfrac{1}{2\sqrt{2}}\ln(v^2+1)$

$\dfrac{1}{\sqrt{2}}\int \dfrac{dv}{v^2+1} = \dfrac{1}{\sqrt{2}}\arctan v$

$\dfrac{1}{2\sqrt{2}}\ln(v^2+1) + \dfrac{1}{\sqrt{2}}\arctan v - \dfrac{1}{2\sqrt{2}}\ln(w^2+1) + \dfrac{1}{\sqrt{2}}\arctan w + c$

$dv = \sqrt{2}\,du \quad w = \sqrt{2}u + 1 \quad w - 1 = \sqrt{2}u \quad dw = \sqrt{2}\,du \quad \dfrac{w-1}{\sqrt{2}} = u$

$= \dfrac{1}{\sqrt{2}}\int \dfrac{v+1}{v^2+1}\,dv - \dfrac{1}{\sqrt{2}}\int \dfrac{w-1}{w^2+1}\,dw$

Step 15: $\int \sqrt{\tan x}\,dx = \dfrac{1}{2\sqrt{2}}\ln\left(\dfrac{v^2+1}{w^2+1}\right) + \dfrac{1}{\sqrt{2}}\arctan v + \dfrac{1}{\sqrt{2}}\arctan w + C$

Step 16:

$$\int \sqrt{\tan x}\,dx = \frac{1}{2\sqrt{2}}\ln\left(\frac{(\sqrt{2}u-1)^2+1}{(\sqrt{2}u+1)^2+1}\right) + \frac{1}{\sqrt{2}}\arctan(\sqrt{2}u-1) + \frac{1}{\sqrt{2}}\arctan(\sqrt{2}u+1) + C$$

$$\int \sqrt{\tan x}\,dx = \frac{1}{2\sqrt{2}}\ln\left(\frac{2u^2-2\sqrt{2}u+1+1}{2u^2+2\sqrt{2}u+1+1}\right) + \frac{1}{\sqrt{2}}\arctan(\sqrt{2}u-1) + \frac{1}{\sqrt{2}}\arctan(\sqrt{2}u+1) + C$$

$$\int \sqrt{\tan x}\,dx = \frac{1}{2\sqrt{2}}\ln\left(\frac{2u^2-2\sqrt{2}u+2}{2u^2+2\sqrt{2}u+2}\right) + \frac{1}{\sqrt{2}}\arctan(\sqrt{2}u-1) + \frac{1}{\sqrt{2}}\arctan(\sqrt{2}u+1) + C$$

$$\int \sqrt{\tan x}\,dx = \frac{1}{2\sqrt{2}}\ln\left(\frac{u^2-\sqrt{2}u+1}{u^2+\sqrt{2}u+1}\right) + \frac{1}{\sqrt{2}}\arctan(\sqrt{2}u-1) + \frac{1}{\sqrt{2}}\arctan(\sqrt{2}u+1) + C$$

$$\int \sqrt{\tan x}\,dx = \frac{1}{2\sqrt{2}}\ln\left(\frac{(\sqrt{\tan x})^2-\sqrt{2}\tan x+1}{(\sqrt{\tan x})^2+\sqrt{2}\tan x+1}\right) +$$
$$\frac{1}{\sqrt{2}}\arctan(\sqrt{2}\tan x-1) + \frac{1}{\sqrt{2}}\arctan(\sqrt{2}\tan x+1) + C$$

$$\int \sqrt{\tan x}\,dx = \frac{1}{2\sqrt{2}}\ln\left(\frac{\tan x-\sqrt{2}\tan x+1}{\tan x+\sqrt{2}\tan x+1}\right) +$$
$$\frac{1}{\sqrt{2}}\arctan(\sqrt{2}\tan x-1) + \frac{1}{\sqrt{2}}\arctan(\sqrt{2}\tan x+1) + C$$

Have students show the answers are equivalent.

**integration by parts** Find the average value of $\ln x$ on the interval $[1, e]$.

Solution: Using $\int u\,dv = uv - \int v\,du$ $\quad \frac{1}{e-1}\int_1^e \ln x\,dx \quad$ u=lnX $\quad$ v=x $\quad du=\frac{1}{x}dx$

dv=dx

$$= \frac{1}{e-1}\left([x\ln x]_1^e - \int_1^e dx\right) \quad \text{since}\left(\frac{1}{x}\cdot x = 1\right) \quad = \frac{1}{e-1}\left(e\ln e - |\ln 1 - x|_1^e\right)$$

(plug in e&1) $= \frac{1}{e-1}(e-0-e+1) = \frac{1}{e-1}$.

basic **partial fractions.** Find $\int \frac{5dx}{x^2+25x+24}$.

Solution: $\int \frac{5}{x^2+25x+24}dx$, factor $\rightarrow \int \frac{5dx}{(x+24)(x+1)} \quad \frac{5}{(x+24)(x+1)} = \frac{A}{x+24} + \frac{B}{x+1}$

$\rightarrow$ 5=(x+1)A+(x+24)B. since this equation must be true for any $x$ pick one that will make one of the terms drop out $x = -1$, $B = \frac{5}{23}$    $x = -24$, $A = -\frac{5}{23}$ $\rightarrow \int \frac{5}{-23(x+24)} dx + \int \frac{5}{23(x+1)} dx = -\frac{5}{23} \ln(x+24) + \frac{5}{23} \ln(x+1) + C = \frac{5}{23} \ln(\frac{x+1}{x+24}) + C$.

advanced **u substitution and partial fractions**. Find $\int \frac{\cos x}{\sin^2 x + 7 \sin x + 10} dx$.

Solution: the solution is straightforward just follow the steps. $\int \frac{\cos x}{\sin^2 x + 7 \sin x + 10} dx$, the integral simplifies to $\int \frac{du}{u^2 + 7u + 10} = \int \frac{du}{(u+2)(u+5)}$   let u=sinx then du=cosx   Partial fractions: 1 = A(u+5)+B(u+2)  Now plug in u values that will make one of the terms drop out.

$u = -5$, $B = -\frac{1}{3}$   $u = -2$, $A = \frac{1}{3}$   $\frac{1}{3}(\int \frac{1}{u+2} du - \int \frac{1}{u+5} du) = \frac{1}{3}(\ln|u+2| - \ln|u+5|) = \frac{1}{3} \ln \left| \frac{\sin x + 2}{\sin x + 5} \right| + C$.

intermediate **two u-substitutions and integration by parts**. Evaluate: $\int (x-5)^7 \sin(x-5)^4 dx$.

Solution: Well the solution is straightforward so just follow the steps $\int (x-5)^7 \sin(x-5)^4 dx$, for this problem we use u=x-5, so it' simply: $\int u^7 \sin u^4 du$ now we use a v substitution, let $v = u^4$ then $dv = 4u^3 du$, now it's $\int \frac{v \sin v}{4} dv$. And now we use integration by parts, let $s = \frac{v}{4}$, and df=sinvdv, ds= $\frac{dv}{4}$, and f=-cosv $\int \frac{v \sin v}{4} dv = -\frac{v}{4} \cos v - \int \frac{-\cos v}{4} dv = -\frac{v}{4} \cos v + \frac{\sin v}{4} + C$   plug back the beginning values for u and v

$= -\frac{(x-5)^4}{4} \cos(x-5)^4 + \frac{\sin(x-5)^4}{4} + C$.

basic **L Hopital s rule** Use L'Hopital's rule to find the limit: $\lim_{x \to \infty} \frac{2^x}{x^2}$

Solution: $\lim_{x \to \infty} \frac{2^x}{x^2}$ since plugging in the limit gives us indeterminate form we can take $= \lim_{x \to \infty} \frac{2^x \ln 2}{2x}$ the derivative of numerator and denominator and take the limit. $= \lim_{x \to \infty} \frac{2^x (\ln 2)^2}{2} = \infty$ repeating l'hopital's rule and evaluate plugging in infinity.

## 7.1 Indefinite Integrals Calculus

**Learning Objectives**

A student will be able to:
- Find antiderivatives of functions.
- Represent antiderivatives.
- Interpret the constant of integration graphically.
- Solve differential equations.
- Use basic anti-differentiation techniques.
- Use basic integration rules.

**Introduction**

In this lesson we will introduce the idea of the *antiderivative* of a function and formalize as *indefinite integrals*. We will derive a set of rules that will aid our computations as we solve problems.

*Antiderivatives*

Definition

A function $F(x)$ is called an *antiderivative* of a function $f$ if $F'(x) = f(x)$ for all $x$ in the domain of $f$.

**Example 1:**

Consider the function $f(x) = 3x^2$. Can you think of a function $F(x)$ such that $F'(x) = f(x)$?
*(Answer: $F(x) = x^3$, $F(x) = x^3 - 6$, many other examples.)*

Since we differentiate $F(x)$ to get $f(x)$, we see that $F(x) = x^3 + C$ will work for any constant $C$. Graphically, we can think the set of all antiderivatives as vertical transformations of the graph of $F(x) = x^3$. The figure shows two such transformations.

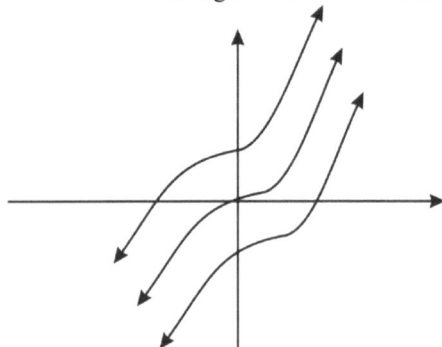

With our definition and initial example, we now look to formalize the definition and develop some useful rules for computational purposes, and begin to see some applications.

*Notation and Introduction to Indefinite Integrals*

The process of finding antiderivatives is called *anti-differentiation*, more commonly referred to as *integration*. We have a particular sign and set of symbols we use to indicate integration:

$$\int f(x)dx = F(x) + C.$$

We refer to the left side of the equation as "the indefinite integral of $f(x)$ with respect to $x$." The function $f(x)$ is called the **integrand** and the constant $C$ is called the **constant of integration.** Finally the symbol $dx$ indicates that we are to integrate with respect to $x$.

Using this notation, we would summarize the last example as follows:

$$\int 3x^2 \, dx = x^3 + C$$

*Using Derivatives to Derive Basic Rules of Integration*

As with differentiation, there are several useful rules that we can derive to aid our computations as we solve problems. The first of these is a rule for integrating power functions, $f(x) = x^n \; [n \neq -1]$, and is stated as follows:

$$\int x^n \, dx = \frac{1}{n+1} x^{n+1} + C.$$

We can easily prove this rule. Let $F(x) = \frac{1}{n+1} x^{n+1} + C, \; n \neq -1$. We differentiate with respect to $x$ and we have:

$$F'(x) = \frac{d}{dx}\left(\frac{1}{n+1} x^{n+1} + C\right) = \frac{d}{dx}\left(\frac{1}{n+1} x^{n+1}\right) + \frac{d}{dx}(C)$$

$$= \left(\frac{1}{n+1}\right) \frac{d}{dx}(x^{n+1}) + \frac{d}{dx}(C)$$

$$= \left(\frac{n+1}{n+1}\right) x^n + 0$$

$$= x^n.$$

The rule holds for $f(x) = x^n [n \neq -1]$. What happens in the case where we have a power function to integrate with $n = -1$, say $\int x^{-1} \, dx = \int \frac{1}{x} dx$. We can see that the rule does not work since it would result in division by $0$. However, if we pose the problem as finding $F(x)$ such that $F'(x) = \frac{1}{x}$, we recall that the derivative of logarithm functions had this form. In particular, $\frac{d}{dx} \ln x = \frac{1}{x}$. Hence

$$\int \frac{1}{x} dx = \ln x + C.$$

In addition to logarithm functions, we recall that the basic exponentional function, $f(x) = e^x$, was special in that its derivative was equal to itself. Hence we have

$$\int e^x \, dx = e^x + C.$$

Again we could easily prove this result by differentiating the right side of the equation above. The actual proof is left as an exercise to the student.

As with differentiation, we can develop several rules for dealing with a finite number of integrable functions. They are stated as follows:

If $f$ and $g$ are integrable functions, and $C$ is a constant, then

$$\int [f(x) + g(x)] \, dx = \int f(x) \, dx + \int g(x) \, dx,$$

$$\int [f(x) - g(x)] \, dx = \int f(x) \, dx - \int g(x) \, dx.$$

$$\int [Cf(x)] \, dx = C \int f(x) \, dx.$$

**Example 2:**
Compute the following indefinite integral.

$$\int \left[2x^3 + \frac{3}{x^2} - \frac{1}{x}\right] dx.$$

**Solution:**
Using our rules we have

$$\int \left[2x^3 + \frac{3}{x^2} - \frac{1}{x}\right] dx = 2\int x^3 dx + 3\int \frac{1}{x^2} dx - \int \frac{1}{x} dx$$

$$= 2\left(\frac{x^4}{4}\right) + 3\left(\frac{x^{-1}}{-1}\right) - \ln x + C$$

$$= \frac{x^4}{2} - \frac{3}{x} - \ln x + C.$$

Sometimes our rules need to be modified slightly due to operations with constants as is the case in the following example.

**Example 3:**
Compute the following indefinite integral:

$$\int e^{3x} dx.$$

**Solution:**
We first note that our rule for integrating exponential functions does not work here since $\frac{d}{dx}e^{3x} = 3e^{3x}$. However, if we remember to divide the original function by the constant then we get the correct antiderivative and have

$$\int e^{3x} dx = \frac{e^{3x}}{3} + C.$$

We can now re-state the rule in a more general form as

$$\int e^{kx} dx = \frac{e^{kx}}{k} + C.$$

***Differential Equations***

We conclude this lesson with some observations about integration of functions. First, recall that the integration process allows us to start with function $f$ from which we find another function $F(x)$ such that $F'(x) = f(x)$. This latter equation is called a ***differential equation.*** This characterization of the basic situation for which integration applies gives rise to a set of equations that will be the focus of the Lesson on The Initial Value Problem.

**Example 4:**
Solve the general differential equation $f'(x) = x^{\frac{2}{3}} + \sqrt{x}$.

**Solution:**
We solve the equation by integrating the right side of the equation and have

$$f(x) = \int f'(x) dx = \int x^{\frac{2}{3}} dx + \int \sqrt{x} dx.$$

We can integrate both terms using the power rule, first noting that $\sqrt{x} = x^{\frac{1}{2}}$, and have

$$f(x) = \int x^{\frac{2}{3}} dx + \int x^{\frac{1}{2}} dx = \frac{3}{5} x^{\frac{5}{3}} + \frac{2}{3} x^{\frac{3}{2}} + C.$$

**Lesson Summary**
1. We learned to find antiderivatives of functions.
2. We learned to represent antiderivatives.
3. We interpreted constant of integration graphically.
4. We solved general differential equations.
5. We used basic antidifferentiation techniques to find integration rules.

6. We used basic integration rules to solve problems.

**Multimedia Link**

The following applet shows a graph, $f(x)$ and its derivative, $f'(x)$. This is similar to other applets we've explored with a function and its derivative graphed side-by-side, but this time $f(x)$ is on the right, and $f'(x)$ is on the left. If you edit the definition of $f'(x)$, you will see the graph of $f(x)$ change as well. The c parameter adds a constant to $f(x)$. Notice that you can change the value of c without affecting $f'(x)$. Why is this?

**Review Questions**

In problems #1–3, find an antiderivative of the function
1. $f(x) = 1 - 3x^2 - 6x$
2. $f(x) = x - x^{\frac{2}{3}}$
3. $f(x) = \sqrt[5]{2x+1}$

In #4–7, find the indefinite integral
4. $\int (2 + \sqrt{5}) dx$
5. $\int 2(x-3)^3 dx$
6. $\int (x^2 \cdot \sqrt[3]{x}) dx$
7. $\int x + \dfrac{1}{x\sqrt[4]{x}} dx$
8. Solve the differential equation $f'(x) = 4x^3 - 3x^2 + x - 3$.
9. Find the antiderivative $F(x)$ of the function $f(x) = 2e^{2x} + x - 2$ that satisfies $F(0) = 5$.
10. Evaluate the indefinite integral $\int |x| dx$. (Hint: Examine the graph of $f(x) = |x|$.)

**Review Answers**

1. $F(x) = x - x^3 - 3x^2 + C$
2. $F(x) = \frac{x^2}{2} - \frac{3}{5} x^{\frac{5}{3}} + C$
3. $F(x) = \frac{5}{12}(2x+1)^{\frac{6}{5}} + C$
4. $\int (2 + \sqrt{5}) dx = 2x + \sqrt{5}x + C$
5. $\int 2(x-3)^3 dx = \frac{(x-3)^4}{2} + C$
6. $\int (x^2 \cdot \sqrt[3]{x}) dx = \frac{3}{10} x^{\frac{10}{3}} + C$
7. $\int x + \dfrac{1}{x\sqrt[4]{x}} dx = \dfrac{x^2}{2} - \dfrac{4}{\sqrt[4]{x}} + C$
8. $f(x) = x^4 - x^3 + \frac{x^2}{2} - 3x + C$
9. $F(x) = e^{2x} + \frac{x^2}{x} - 2x + 4$
10. $\int |x| dx = \frac{x^2}{2} + C$

# Indefinite Integrals Practice

1. Verify the statement by showing that the derivative of the right side is equal to the integrand of the left side.

   a. $\displaystyle\int\left(-\frac{9}{x^4}\right)dx = \frac{3}{x^3}+C$

   b. $\displaystyle\int\left(1-\frac{1}{\sqrt[3]{x^2}}\right)dx = x-3\sqrt[3]{x}+C$

2. Integrate.

   a. $\displaystyle\int 6\,dx$

   b. $\displaystyle\int 3t^2\,dt$

   c. $\displaystyle\int 5x^{-3}\,dx$

   d. $\displaystyle\int du$

   e. $\displaystyle\int x^{3/2}\,dx$

   f. $\displaystyle\int \sqrt[3]{x}\,dx$

   g. $\displaystyle\int \frac{1}{x\sqrt{x}}\,dx$

   h. $\displaystyle\int \frac{1}{2x^3}\,dx$

   i. $\displaystyle\int (x^3+2)\,dx$

   j. $\displaystyle\int \left(2x^{4/3}+3x-1\right)dx$

   k. $\displaystyle\int \sqrt[3]{x^2}\,dx$

   l. $\displaystyle\int \frac{1}{x^3}\,dx$

   m. $\displaystyle\int \frac{1}{4x^2}\,dx$

   n. $\displaystyle\int \frac{t^2+2}{t^2}\,dt$

   o. $\displaystyle\int u(3u^2+1)\,du$

   p. $\displaystyle\int (x-1)(6x-5)\,dx$

   q. $\displaystyle\int y^2\sqrt{y}\,dy$

3. Find two functions that have the given derivative and sketch the graph of each.

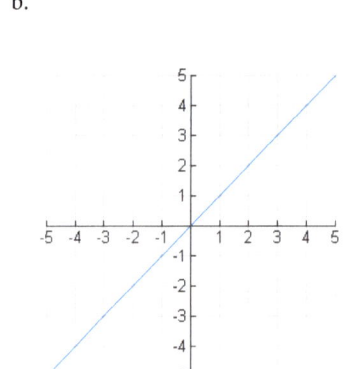

Answers: (Of course, you could have checked all of yours using differentiation!)

2. a. $6x+C$  b. $t^3+C$  c. $-\dfrac{5}{2x^2}+C$  d. $u+C$

   e. $\dfrac{2}{5}x^{5/2}+C$  f. $\dfrac{3}{4}\sqrt[3]{x^4}+C$  g. $\dfrac{-2}{\sqrt{x}}+C$  h. $-\dfrac{1}{4x^2}+C$

i. $\dfrac{x^4}{4}+2x+C$  
j. $\dfrac{6}{7}x^{7/3}+\dfrac{3}{2}x^2-x+C$  
k. $\dfrac{3}{5}x^{5/3}+C$  
l. $-\dfrac{1}{2x^2}+C$  
m. $-\dfrac{1}{4x}+C$  
n. $t-\dfrac{2}{t}+C$  
o. $\dfrac{3}{4}u^4+\dfrac{1}{2}u^2+C$  
p. $2x^3-\dfrac{11}{2}x^2+5x+C$  
q. $\dfrac{2}{7}y^{7/2}+C$

## 7.2 The Initial Value Problem

**Learning Objectives**
- Find general solutions of differential equations
- Use initial conditions to find particular solutions of differential equations

**Introduction**

In the Lesson on Indefinite Integrals Calculus we discussed how finding antiderivatives can be thought of as finding solutions to differential equations: $F'(x) = f(x)$. We now look to extend this discussion by looking at how we can designate and find particular solutions to differential equations. Let's recall that a general differential equation will have an infinite number of solutions. We will look at one such equation and see how we can impose conditions that will specify exactly one particular solution.

**Example 1:**

Suppose we wish to solve the following equation:
$f'(x) = e^{3x} - 6\sqrt{x}.$

**Solution:**

We can solve the equation by integration and we have
$f(x) = \frac{1}{3}e^{3x} - 4x^{\frac{3}{2}} + C.$

We note that there are an infinite number of solutions. In some applications, we would like to designate exactly one solution. In order to do so, we need to impose a condition on the function $f$. We can do this by specifying the value of $f$ for a particular value of $x$. In this problem, suppose that add the condition that $f(0) = 1$. This will specify exactly one value of $C$ and thus one particular solution of the original equation:

Substituting $f(0) = 1$ into our general solution $f(x) = \frac{1}{3}e^{3x} - 4x^{\frac{3}{2}} + C$ gives $1 = \frac{1}{3}e^{3(0)} - 4(0)^{\frac{3}{2}} + C$ or $C = 1 - \frac{1}{3} = \frac{2}{3}$. Hence the solution $f(x) = \frac{1}{3}e^{3x} - 4x^{\frac{3}{2}} + \frac{2}{3}$ is the ***particular solution*** of the original equation $f'(x) = e^{3x} - 6\sqrt{x}$ satisfying the ***initial condition*** $f(0) = 1$.

We now can think of other problems that can be stated as differential equations with initial conditions. Consider the following example.

**Example 2:**

Suppose the graph of $f$ includes the point $(2, 6)$ and that the slope of the tangent line to $f$ at any point $x$ is given by the expression $3x + 4$. Find $f(-2)$.

**Solution:**

We can re-state the problem in terms of a differential equation that satisfies an initial condition.
$f'(x) = 3x + 4$ with $f(2) = 6$.

By integrating the right side of the differential equation we have

$f(x) = \frac{3}{2}x^2 + 4x + C$ as the general solution. Substituting the condition that $f(2) = 6$ gives

$6 = \frac{3}{2}(2)^2 + 4(2) + C.$

$6 = 6 + 8 + C.$

$C = -8.$

Hence $f(x) = \frac{3}{2}x^2 + 4x - 8$ is the ***particular solution*** of the original equation $f'(x) = 3x + 4$ satisfying the ***initial condition*** $f(2) = 6$.

Finally, since we are interested in the value $f(-2)$, we put $-2$ into our expression for $f$ and obtain:
$f(-2) = -10$

**Lesson Summary**

1. We found general solutions of differential equations.
2. We used initial conditions to find particular solutions of differential equations.

**Multimedia Link**

The following applet allows you to set the initial equation for $f'(x)$ and then the slope field for that equation is displayed. In magenta you'll see one possible solution for $f(x)$. If you move the magenta point to the initial value, then you will see the graph of the solution to the initial value problem. Follow the directions on the page with the applet to explore this idea, and then try redoing the examples from this section on the apple.

**Review Questions**

In problems #1–3, solve the differential equation for $f(x)$.
1. $f'(x) = 2e^{2x} - 2\sqrt{x}$
2. $f'(x) = \sin x - \frac{1}{e^x}$
3. $f''(x) = (2+x)\sqrt{x}$

In problems #4–7, solve the differential equation for $f(x)$ given the initial condition.
4. $f'(x) = 6x^5 - 4x^2 + \frac{7}{3}$ and $f(1) = 4$.
5. $f'(x) = 3x^2 + e^{2x}$ and $f(0) = 3$.
6. $f'(x) = \sqrt[3]{x^2} - \frac{1}{x^2}$ and $f(1) = 3$
7. $f'(x) = (2\cos x - \sin x), -\frac{\pi}{2} \leq x \leq \frac{\pi}{2}$, and $f(\frac{\pi}{3}) = \sqrt{3} + \frac{1}{2}$
8. Suppose the graph of f includes the point (-2, 4) and that the slope of the tangent line to f at x is -2x+4. Find f(5).

In problems #9–10, find the function $f$ that satisfies the given conditions.
9. $f''(x) = \sin x - e^{-2x}$ with $f'(0) = \frac{5}{2}$ and $f(0) = 0$
10. $f''(x) = \frac{1}{\sqrt{x}}$ with $f'(4) = 7$ and $f(4) = 25$

**Review Answers**
1. $f(x) = 2^{2x} - \frac{4}{3}x^{\frac{3}{2}} + C$
2. $f(x) = -\cos x + \frac{1}{e^x} + C$
3. $f(x) = \frac{8}{15}x^{\frac{5}{2}} + \frac{4}{35}x^{\frac{7}{2}} + C$
4. $f(x) = x^6 - \frac{4}{3}x^3 + \frac{7}{3}x + 2$
5. $f(x) = x^3 + \frac{e^{2x}}{2} + \frac{5}{2}$
6. $f(x) = \frac{3}{5}\sqrt[3]{x^5} + \frac{1}{x} + \frac{7}{5}$
7. $f(x) = 2\sin x + \cos x$
8. $f(x) = -x^2 + 4x + 16$; $f(5) = 11$
9. $f(x) = -\sin x - \frac{1}{4}e^{2x} + 4x + \frac{1}{4}$
10. $f(x) = \frac{4}{3}x^{\frac{3}{2}} + 3x + \frac{7}{3}$

# Initial Condition & Integration of Trig Functions Practice

1. Find the particular solution $y = f(x)$ that satisfies the differential equation and initial condition.

   a. $f'(x) = 3\sqrt{x} + 3,\ f(1) = 4$

   b. $f'(x) = 6x(x-1),\ f(10) = -10$

   c. $f'(x) = \dfrac{2-x}{x^3},\ x > 0,\ f(2) = \dfrac{3}{4}$

   d. $f'(x) = \sec^2 x,\ f\left(\dfrac{\pi}{3}\right) = 2\sqrt{3}$

2. Find the equation of the function $f$ whose graph passes through the point.

   $f'(x) = 6\sqrt{x} - 10,\ (4, 2)$

3. Find the function $f$ that satisfies the given conditions.

   a. $f''(x) = 2,\ f'(2) = 5,\ f(2) = 10$

   b. $f''(x) = x^{-2/3},\ f'(8) = 6,\ f(0) = 0$

4. Integrate.

   a. $\displaystyle\int (2\sin x + 3\cos x)\,dx$

   b. $\displaystyle\int (1 - \csc t \cot t)\,dt$

   c. $\displaystyle\int (\csc^2 \theta - \cos\theta)\,d\theta$

   d. $\displaystyle\int (t^2 - \sin t)\,dt$

Answers:

1. 
   a. $f(x) = 2x^{3/2} + 3x - 1$
   
   b. $f(x) = 2x^3 - 3x^2 - 1710$
   
   c. $f(x) = \dfrac{-1}{x^2} + \dfrac{1}{x} + \dfrac{1}{2}$
   
   d. $f(x) = \tan x + \sqrt{3}$

2. $f(x) = 4x^{3/2} - 10x + 10$

3. 
   a. $f(x) = x^2 + x + 4$
   
   b. $f(x) = \dfrac{9}{4}x^{4/3}$

4. 
   a. $-2\cos x + 3\sin x + C$
   
   b. $t + \csc t + C$
   
   c. $-\cot\theta - \sin\theta + C$
   
   d. $\dfrac{t^3}{3} + \cos t + C$

## 7.3 The Area Problem

**Learning Objectives**
- Use sigma notation to evaluate sums of rectangular areas
- Find limits of upper and lower sums
- Use the limit definition of area to solve problems

**Introduction**

In The Lesson The Calculus we introduced the area problem that we consider in integral calculus. The basic problem was this:

$f(x) = x^2$. Suppose we are interested in finding the area between the $x$-axis and the curve of $f(x) = x^2$ from $x = 0$ to $x = 1$.

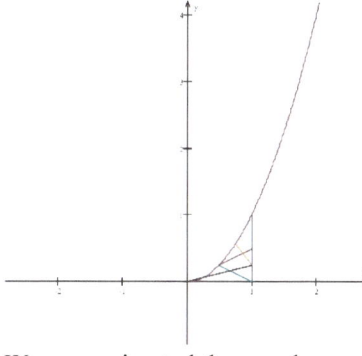

We approximated the area by constructing four rectangles, with the height of each rectangle equal to the maximum value of the function in the sub-interval.

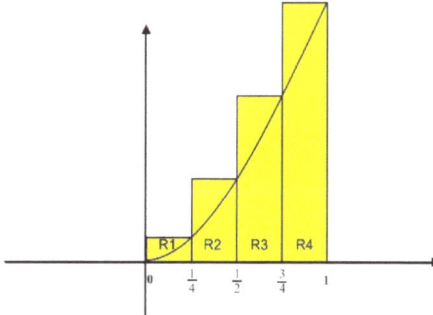

We then summed the areas of the rectangles as follows:

and $R_1 + R_2 + R_3 + R_4 = \frac{30}{64} = \frac{15}{32} \approx 0.46$.

We call this the **upper sum** since it is based on taking the maximum value of the function within each sub-interval. We noted that as we used more rectangles, our area approximation became more accurate.

We would like to formalize this approach for both upper and lower sums. First we note that the **lower sums** of the area of the rectangles results in $R_1 + R_2 + R_3 + R_4 = 13/64 \approx 0.20$. Our intuition tells us that the true area lies somewhere between these two sums, or $0.20 < \text{Area} < 0.46$ and that we will get closer to it by using more and more rectangles in our approximation scheme.

In order to formalize the use of sums to compute areas, we will need some additional notation and terminology.

**Sigma Notation**

In The Lesson The Calculus we used a notation to indicate the upper sum when we increased our rectangles to $N = 16$ and found that our approximation $A = \sum_1^{16} R_i = \frac{195}{512} \approx 0.38$. The notation we used to enabled us to indicate the sum without the need to write out all of the individual terms. We will make use of this notation as we develop more formal definitions of the area under the curve.

Let's be more precise with the notation. For example, the quantity $A = \sum R_i$ was found by summing the areas of $N = 16$ rectangles. We want to indicate this process, and we can do so by providing indices to the symbols used as follows:

$$A = \sum_{i=1}^{16} R_i = R_1 + R_2 + R_3 + \ldots + R_{15} + R_{16}.$$

The sigma symbol with these indices tells us how the rectangles are labeled and how many terms are in the sum.

## Useful Summation Formulas

We can use the notation to indicate useful formulas that we will have occasion to use. For example, you may recall that the sum of the first $n$ integers is $n(n+1)/2$. We can indicate this formula using sigma notation. The formula is given here along with two other formulas that will become useful to us.

$$\sum_{i=1}^{n} i = \frac{n(n+1)}{2},$$

$$\sum_{i=1}^{n} i^2 = \frac{n(n+1)(2n+1)}{6},$$

$$\sum_{i=1}^{n} i^3 = \left[\frac{n(n+1)}{2}\right]^2.$$

We can show from associative, commutative, and distributive laws for real numbers that
$\sum_{i=1}^{n}(a_i + b_i) = \sum_{i=1}^{n}(a_i) + \sum_{i=1}^{n}(b_i)$ and
$\sum_{i=1}^{n}(ka_i) = k\sum_{i=1}^{n}(a_i)$.

**Example 1:**

Compute the following quantity using the summation formulas:

$$\sum_{i=1}^{10} 2i(i - 6i).$$

**Solution:**

$$\sum_{i=1}^{10} 2i(i - 6i) = \sum_{i=1}^{10}(2i^2 - 12i) = 2\sum_{i=1}^{10} i^2 - 12\sum_{i=1}^{10} i$$

$$= 2\left(\frac{(10)(10+1)(2 \cdot 10 + 1)}{6}\right) - 12\left(\frac{(10)(11)}{2}\right)$$

$$= 770 - 660 = 110.$$

## Another Look at Upper and Lower Sums

We are now ready to formalize our initial ideas about upper and lower sums.

Let $f$ be a bounded function in a closed interval $[a, b]$ and $P = [x_0, \ldots, x_n]$ the partition of $[a, b]$ into $n$ subintervals.

We can then define the lower and upper sums, respectively, over partition $P$, by

$$S(P) = \sum_{1}^{n} m_i(x_i - x_{i-1}) = m_1(x_1 - x_0) + m_2(x_2 - x_1) + \ldots + m_n(x_n - x_{n-1}).$$

$$T(P) = \sum_{1}^{n} M_i(x_i - x_{i-1}) = M_1(x_1 - x_0) + M_2(x_2 - x_1) + \ldots + M_n(x_n - x_{n-1}).$$

where $m_i$ is the minimum value of $f$ in the interval of length $x_i - x_{i-1}$ and $M_i$ is the maximum value of $f$ in the interval of length $x_i - x_{i-1}$.

The following example shows how we can use these to find the area.

**Example 2:**

Show that the upper and lower sums for the function $f(x) = x^2$ from $x = 0$ to $x = 1$ approach the value $A = 1/3$.

**Solution:**

Let $P$ be a partition of $n$ equal sub intervals over $[0, 1]$. We will show the result for the upper sums. By our definition we have

$$T(P) = \sum_1^n M_i(x_i - x_{i-1}) = M_1(x_1 - x_0) + M_2(x_2 - x_1) + \ldots + M_n(x_n - x_{n-1}).$$

We note that each rectangle will have width $\frac{1}{n}$, and lengths $\left(\frac{1}{n}\right)^2, \left(\frac{2}{n}\right)^2, \left(\frac{3}{n}\right)^2, \ldots, \left(\frac{n}{n}\right)^2$ as indicated:

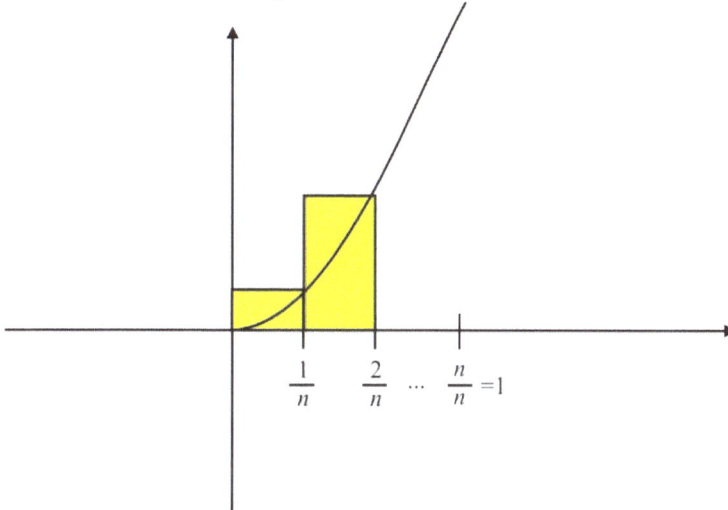

$$T(P) = \sum_1^n M_i(x_i - x_{i-1}) = M_1(x_1 - x_0) + M_2(x_2 - x_1) + \ldots + M_n(x_n - x_{n-1})$$

$$= \frac{1}{n}\left(\frac{1}{n}\right)^2 + \frac{1}{n}\left(\frac{2}{n}\right)^2 + \frac{1}{n}\left(\frac{3}{n}\right)^2 + \ldots + \frac{1}{n}\left(\frac{n}{n}\right)^2$$

$$= \frac{1}{n}\left(\frac{1}{n}\right)^2 (1^2 + 2^2 + 3^2 + \ldots + n^2)$$

$$= \left(\frac{1}{n^3}\right)(1^2 + 2^2 + 3^2 + \ldots + n^2) = \left(\frac{1}{n^3}\right)\left(\frac{n(n+1)(2n+1)}{6}\right) = \left(\frac{(n+1)(2n+1)}{6n^2}\right).$$

We can re-write this result as:

$$\frac{(n+1)(2n+1)}{6n^2} = \frac{1}{6}\left(\frac{n+1}{n}\right)\left(\frac{2n+1}{n}\right) = \frac{1}{6}\left(1 + \frac{1}{n}\right)\left(2 + \frac{1}{n}\right).$$

We observe that as

$$x \to +\infty, \frac{1}{6}\left(1 + \frac{1}{n}\right)\left(2 + \frac{1}{n}\right) \to \frac{1}{3}.$$

We now are able to define the area under a curve as a limit.

**Definition**

Let $f$ be a continuous function on a closed interval $[a, b]$. Let $P$ be a partition of $n$ equal sub intervals over $[a, b]$. Then the area under the curve of $f$ is the limit of the upper and lower sums, that is
$$A = \lim_{n \to +\infty} S(P) = \lim_{n \to +\infty} T(P).$$

**Example 3:**

Use the limit definition of area to find the area under the function $f(x) = 4 - x$ from 1 to $x = 3$.

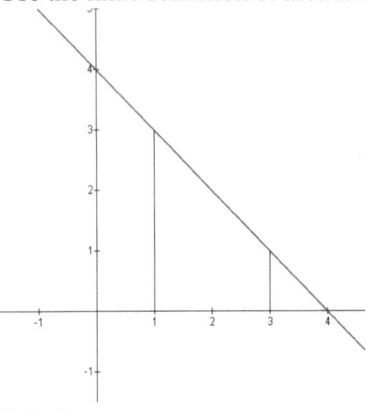

**Solution:**

If we partition the interval $[1, 3]$ into $n$ equal sub-intervals, then each sub-interval will have length $\frac{3-1}{n} = \frac{2}{n}$ and height $3 - i\triangle x$ as $i$ varies from 1 to $n$. So we have $\triangle x = \frac{2}{n}$ and

$$S(P) = \sum_1^n (3 - i\triangle x)\triangle x = \sum_1^n (3\triangle x) - \sum_1^n i(\triangle x)^2$$
$$= (3\triangle x)n - \frac{n(n+1)}{n}(\triangle x)^2.$$

Since $\triangle x = \frac{2}{n}$, we then have by substitution as $n \to \infty$. Hence the area is $A = 4$.

This example may also be solved with simple geometry. It is left to the reader to confirm that the two methods yield the same area.

**Lesson Summary**
1. We used sigma notation to evaluate sums of rectangular areas.
2. We found limits of upper and lower sums.
3. We used the limit definition of area to solve problems.

**Review Questions**

In problems #1–2, find the summations.
1. $\sum_{i=1}^{10} i(2i - 3)$
2. $\sum_{i=1}^{n} (3-i)(2+i)$

In problems #3–5, find $S(P)$ and $T(P)$ under the partition P.
3. $f(x) = 1 - x^2, P = \{0, \frac{1}{2}, 1, \frac{3}{2}, 2\}$
4. $f(x) = 2x^2, P = \{-1, -\frac{1}{2}, 0, \frac{1}{2}, 1\}$
5. $f(x) = \frac{1}{x}, P = \{-4, -3, -2, -1\}$

In problems #6–8, find the area under the curve using the limit definition of area.
6. $f(x) = 3x + 5$ from $x = 2$ to $x = 6$.
7. $f(x) = x^2$ from $x = 1$ to $x = 3$.
8. $f(x) = \frac{1}{x}$ from $x = 1$ to $x = 4$.

In problems #9–10, state whether the function is integrable in the given interval. Give a reason for your answer.

9. $f(x) = |x - 2|$ on the interval $[1, 4]$

10. $f(x) = \begin{cases} 1 & \text{if } x \text{ is rational} \\ -1 & \text{if } x \text{ is irrational} \end{cases}$ on the interval $[0, 1]$

**Review Answers**

1. $\sum_{i=1}^{10} i(2i - 3) = 605$
2. $\sum_{i=1}^{n} (3 - i)(2 + i) = \frac{1}{3}(19 - n^2)$
3. $S(P) = -1.75, T(P) = 0.25$

(note that we have included areas under the x-axis as negative values.)

4. $S(P) = 0.5, T(P) = 2.5$
5. $S(P) = -1.83, T(P) = -1.08$
6. Area $= 68$
7. Area $= \frac{26}{3}$
8. Area $= \frac{15}{16}$
9. Yes, since $f(x) = |x - 2|$ is continuous on $[1, 4]$
10. No, since $S(P) = -1, T(P) = 1$

**7.4 Definite Integrals**

**Learning Objectives**
- Use Riemann Sums to approximate areas under curves
- Evaluate definite integrals as limits of Riemann Sums

**Introduction**

In the Lesson The Area Problem we defined the area under a curve in terms of a limit of sums.

$$A = \lim_{n \to +\infty} S(P) = \lim_{n \to +\infty} T(P)$$

where

$$S(P) = \sum_{1}^{n} m_i(x_i - x_{i-1}) = m_1(x_1 - x_0) + m_2(x_2 - x_1) + \ldots + m_n(x_n - x_{n-1}),$$

$$T(P) = \sum_{1}^{n} M_i(x_i - x_{i-1}) = M_1(x_1 - x_0) + M_2(x_2 - x_1) + \ldots + M_n(x_n - x_{n-1}),$$

$S(P)$ and $T(P)$ were examples of **Riemann Sums**. In general, Riemann Sums are of form $\sum_{i=1}^{n} f(x_i^*) \triangle x$ where each $x_i^*$ is the value we use to find the length of the rectangle in the $i^{th}$ sub-interval. For example, we used the maximum function value in each sub-interval to find the upper sums and the minimum function in each sub-interval to find the lower sums. But since the function is continuous, we could have used any points within the sub-intervals to find the limit. Hence we can define the most general situation as follows:

**Definition**

If $f$ is continuous on $[a, b]$, we divide the interval $[a, b]$ into $n$ sub-intervals of equal width with $\triangle x = \frac{b-a}{n}$. We let $x_0 = a, x_1, x_2, \ldots, x_n = b$ be the endpoints of these sub-intervals and let $x_1^*, x_2^*, \ldots, x_n^*$ be *any* sample points in these sub-intervals. Then the **definite integral** of $f$ from $x = a$ to $x = b$ is

$$\int_a^b f(x) dx = \lim_{n \to \infty} \sum_{i=1}^{n} f(x_i^*) \triangle x.$$

**Example 1:**

Evaluate the Riemann Sum for $f(x) = x^3$ from $x = 0$ to $x = 3$ using $n = 6$ sub-intervals and taking the sample points to be the midpoints of the sub-intervals.

**Solution:**

If we partition the interval $[0, 3]$ into $n = 6$ equal sub-intervals, then each sub-interval will have length $\frac{3-0}{6} = \frac{1}{2}$. So we have $\triangle x = \frac{1}{2}$ and

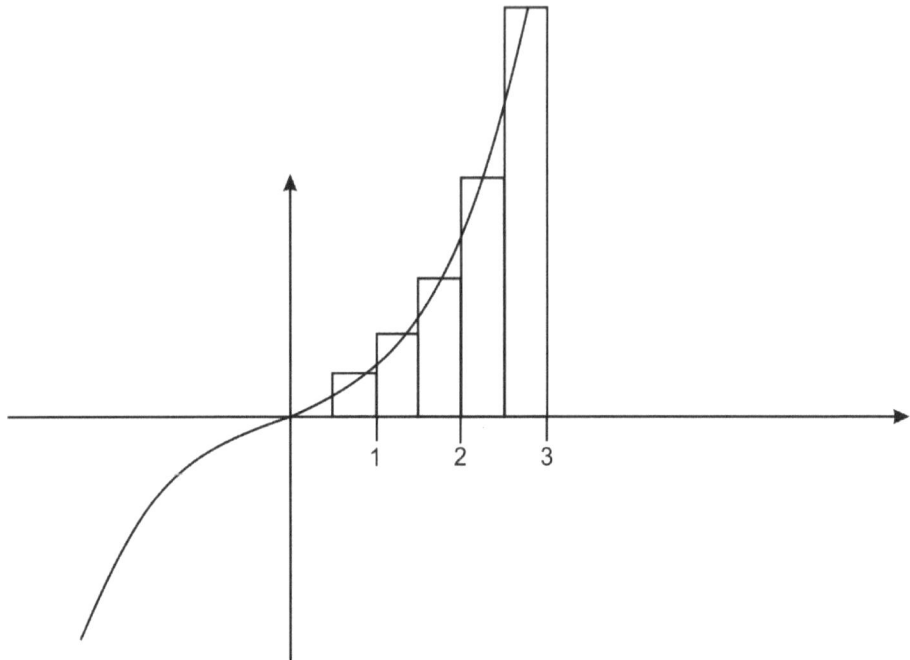

$$R_6 = \sum_{1}^{6} f(x_i^*)\triangle x = f(0.25)\triangle x + f(0.75)\triangle x + f(1.25)\triangle x + f(1.75)\triangle x + f(2.25)\triangle x + f(2.75)\triangle x$$

$$= \left(\frac{1}{64}\right)\left(\frac{1}{2}\right) + \left(\frac{27}{64}\right)\left(\frac{1}{2}\right) + \left(\frac{125}{64}\right)\left(\frac{1}{2}\right) + \left(\frac{343}{64}\right)\left(\frac{1}{2}\right) + \left(\frac{729}{64}\right)\left(\frac{1}{2}\right) + \left(\frac{1331}{64}\right)\left(\frac{1}{2}\right)$$

$$= \frac{2556}{64} = 39.93.$$

Now let's compute the definite integral using our definition and also some of our summation formulas.

**Example 2:**

Use the definition of the definite integral to evaluate $\int_0^3 x^3 dx$.

**Solution:**

Applying our definition, we need to find

$$\int_0^3 x^3 dx = \lim_{n \to \infty} \sum_{i=1}^{n} f(x_i^*)\triangle x.$$

We will use right endpoints to compute the integral. We first need to divide $[0, 3]$ into $n$ sub-intervals of length $\triangle x = \frac{3-0}{n} = \frac{3}{n}$. Since we are using right endpoints, $x_0 = 0, x_1 = \frac{3}{n}, x_2 = \frac{6}{n}, \ldots x_i = \frac{3i}{n}$.

So $\int_0^3 x^3 dx =$

Recall that $\sum_1^n i^3 = \left[\frac{n(n+1)}{2}\right]^2$. By substitution, we have

$$\int_0^3 x^3 dx = \lim_{n\to\infty} \frac{81}{n^4}\left[\frac{n(n+1)}{2}\right]^2 = \lim_{n\to\infty} \frac{81}{4}\left[1+\frac{1}{n}\right]^2 \to \frac{81}{4} \text{ as } n \to \infty.$$

Hence

$$\int_0^3 x^3 dx = \frac{81}{4}.$$

Before we look to try some problems, let's make a couple of observations. First, we will soon not need to rely on the summation formula and Riemann Sums for actual computation of definite integrals. We will develop several computational strategies in order to solve a variety of problems that come up. Second, the idea of definite integrals as approximating the area under a curve can be a bit confusing since we may sometimes get results that do not make sense when interpreted as areas. For example, if we were to compute the definite integral $\int_{-3}^3 x^3 dx$ then due to the symmetry of $f(x) = x^3$ about the origin, we would find that $\int_{-3}^3 x^3 dx = 0$. This is because for every sample point $x_j^*$ we also have $-x_j^*$ is also a sample point with $f(-x_j^*) = -f(x_j^*)$. Hence, it is more accurate to say that $\int_{-3}^3 x^3 dx$ gives us the **net area** between $x = -3$ and $x = 3$. If we wanted the **total area** bounded by the graph and the $x$-axis, then we would compute $2\int_0^3 x^3 dx = \frac{81}{2}$.

**Lesson Summary**
1. We used Riemann Sums to approximate areas under curves.

**Review Questions**

In problems #1–7, use Riemann Sums to approximate the areas under the curves.

1. Consider $f(x) = 2 - x$ from $x = 0$ to $x = 2$. Use Riemann Sums with four subintervals of equal lengths. Choose the midpoints of each subinterval as the sample points.
2. Repeat problem #1 using geometry to calculate the exact area of the region under the graph of $f(x) = 2 - x$ from $x = 0$ to $x = 2$. (Hint: Sketch a graph of the region and see if you can compute its area using area measurement formulas from geometry.)
3. Repeat problem #1 using the definition of the definite integral to calculate the exact area of the region under the graph of $f(x) = 2 - x$ from $x = 0$ to $x = 2$.
4. $f(x) = x^2 - x$ from $x = 1$ to $x = 4$. Use Riemann Sums with five subintervals of equal lengths. Choose the left endpoint of each subinterval as the sample points.
5. Repeat problem #4 using the definition of the definite integral to calculate the exact area of the region under the graph of $f(x) = x^2 - x$ from $x = 1$ to $x = 4$.
6. Consider $f(x) = 3x^2$. Compute the Riemann Sum of f on [0, 1] under each of the following situations. In each case, use the right endpoint as the sample points.
    a. Two sub-intervals of equal length.
    b. Five sub-intervals of equal length.
    c. Ten sub-intervals of equal length.
    d. Based on your answers above, try to guess the exact area under the graph of f on [0, 1].
7. Consider $f(x) = e^x$. Compute the Riemann Sum of f on [0, 1] under each of the following situations. In each case, use the right endpoint as the sample points.
    a. Two sub-intervals of equal length.
    b. Five sub-intervals of equal length.
    c. Ten sub-intervals of equal length.
    d. Based on your answers above, try to guess the exact area under the graph of f on [0, 1].

8. Find the net area under the graph of $f(x) = x^3 - x$; $x = -1$ to $x = 1$. (Hint: Sketch the graph and check for symmetry.)
9. Find the total area bounded by the graph of $f(x) = x^3 - x$ and the x-axis, from $x = -1$ to $x = 1$.
10. Use your knowledge of geometry to evaluate the following definite integral: $\int_0^3 \sqrt{9-x^2}\,dx$

    (Hint: set $y = \sqrt{9-x^2}$ and square both sides to see if you can recognize the region from geometry.)

**Review Answers**
1. Area = 2
2. Area = 2
3. Area = 2
4. Area = 10.08
5. Area = 15.5
6. 
    a. Area = 1.875
    b. Area = 1.32
    c. Area = 1.15
    d. Area = 1
7. 
    a. Area = 2.18
    b. Area = 1.89
    c. Area = 1.80
    d. Area = $e^1 - 1 \approx 1.71$
8. The graph is symmetric about the origin; hence net Area = 0
9. Area = 1/2
10. The graph is that of a quarter circle of radius 3; hence Area = $\dfrac{9\pi}{4}$

## 7.5 Evaluating Definite Integrals

**Learning Objectives**
- Use antiderivatives to evaluate definite integrals
- Use the Mean Value Theorem for integrals to solve problems
- Use general rules of integrals to solve problems

**Introduction**

In the Lesson on Definite Integrals, we evaluated definite integrals using the limit definition. This process was long and tedious. In this lesson we will learn some practical ways to evaluate definite integrals. We begin with a theorem that provides an easier method for evaluating definite integrals. Newton discovered this method that uses antiderivatives to calculate definite integrals.

**Theorem:**

If $f$ is continuous on the closed interval $[a, b]$, then

$$\int_a^b f(x)\,dx = F(b) - F(a),$$

where $F$ is any antiderivative of $f$.

We sometimes use the following shorthand notation to indicate $\int_a^b f(x)\,dx = F(b) - F(a)$:

$$\int_a^b f(x)\,dx = F(x)\Big]_a^b.$$

The proof of this theorem is included at the end of this lesson. Theorem 4.1 is usually stated as a part of the Fundamental Theorem of Calculus, a theorem that we will present in the Lesson on the Fundamental Theorem of Calculus. For now, the result provides a useful and efficient way to compute definite integrals. We need only find an antiderivative of the given function in order to compute its integral over the closed interval. It also gives us a result with which we can now state and prove a version of the Mean Value Theorem for integrals. But first let's look at a couple of examples.

**Example 1:**
Compute the following definite integral:
$$\int_0^3 x^3 dx.$$

**Solution:**

Using the limit definition we found that $\int_0^3 x^3 dx = \frac{81}{4}$. We now can verify this using the theorem as follows:

We first note that $x^4/4$ is an antiderivative of $f(x) = x^3$. Hence we have
$$\int_0^3 x^3 dx = \frac{x^4}{4}\Big]_0^3 = \frac{81}{4} - \frac{0}{4} = \frac{81}{4}.$$

We conclude the lesson by stating the rules for definite integrals, most of which parallel the rules we stated for the general indefinite integrals.

$$\int_a^a f(x)dx = 0$$

$$\int_a^b f(x)dx = -\int_b^a f(x)dx$$

$$\int_a^b k \cdot f(x)dx = k\int_a^b f(x)dx$$

$$\int_a^b [f(x) \pm g(x)]dx = \int_a^b f(x)dx \pm \int_a^b g(x)dx$$

$$\int_a^b f(x)dx = \int_a^c f(x)dx + \int_c^b f(x)dx \text{ where } a < c < b.$$

Given these rules together with Theorem 4.1, we will be able to solve a great variety of definite integrals.

**Example 2:**
Compute $\int_{-2}^2 (x - \sqrt{x})dx$.
**Solution:**

$$\int_1^4 (x - \sqrt{x})dx = \int_1^4 x\,dx - \int_1^4 \sqrt{x}\,dx = \frac{x^2}{2}\Big]_1^4 - \frac{2}{3}x^{\frac{3}{2}}\Big]_1^4 = \left(8 - \frac{1}{2}\right) - \frac{2}{3}(8 - 1) = \frac{15}{2} - \frac{14}{3} = \frac{17}{6}.$$

**Example 3:**
Compute $\int_0^{\frac{\pi}{2}} (x + \cos x)dx$.
**Solution:**

$$\int_0^{\frac{\pi}{2}} (x + \cos x)dx = \int_0^{\frac{\pi}{2}} (x)dx + \int_0^{\frac{\pi}{2}} (\cos x)dx = \frac{x^2}{2}\Big]_0^{\frac{\pi}{2}} + \frac{\sin x}{1}\Big]_0^{\frac{\pi}{2}} = \frac{\pi^2}{4} + 1 = \frac{\pi^2 + 4}{4}.$$

**Lesson Summary**
1. We used antiderivatives to evaluate definite integrals.
2. We used the Mean Value Theorem for integrals to solve problems.
3. We used general rules of integrals to solve problems.

**Proof of Theorem 4.1**

We first need to divide $[a,b]$ into $n$ sub-intervals of length $\Delta x = \frac{b-a}{n}$. We let $x_0 = a, x_1, x_2, \ldots, x_n = b$ be the endpoints of these sub-intervals.

Let $F$ be any antiderivative of $f$.

Consider $F(b) - F(a) = F(x_n) - F(x_0)$.

We will now employ a method that will express the right side of this equation as a Riemann Sum. In particular,

$$F(b) - F(a) = F(x_n) - F(x_0)$$
$$= F(x_n) - F(x_{n-1}) + F(x_{n+1}) - F(x_{n-2}) + F(x_{n-2}) - \ldots - F(x_1) + F(x_1)$$
$$= \sum_1^n [F(x_i) - F(x_{i-1})].$$

Note that $F$ is continuous. Hence, by the Mean Value Theorem, there exist $c_i \in [x_i - 1, x_i]$ such that $F(x_i) - F(x_{i-1}) = F'(c_i)(x_i - x_{i-1}) = f(c_i)\Delta x$.

Hence

$$F(b) - F(a) = \sum_1^n F'(c_i)(x_i - x_{i-1}) = \sum_1^n f(c_i)\Delta x.$$

Taking the limit of each side as $n \to \infty$ we have

$$\lim_{n \to \infty}[F(b) - F(a)] = \lim_{n \to \infty} \sum_1^n f(c_i)\Delta x.$$

We note that the left side is a constant and the right side is our definition for $\int_a^b f(x)dx$.

Hence

$$F(b) - F(a) = \lim_{n \to \infty} \sum_1^n f(c_i)\Delta x = \int_a^b f(x)dx.$$

**Proof of Theorem 4.2**

Let $F(x) = \int_a^x f(x)dx$.

By the Mean Value Theorem for derivatives, there exists $c \in [a,b]$ such that

$$F'(c) = \frac{F(b) - F(a)}{b - a}.$$

From Theorem 4.1 we have that $F$ is an antiderivative of $f$. Hence, $F'(x) = f(x)$ and in particular, $F'(c) = f(c)$. Hence, by substitution we have $f(c) = \frac{F(b) - F(a)}{b - a}$.

Note that $F(a) = \int_a^a f(x)dx = 0$. Hence we have $f(c) = \frac{F(b) - 0}{b - a} = \frac{F(b)}{b - a}$,

and by our definition of $F(x)$ we have $f(c) = \frac{1}{b-a}F(b) = \frac{1}{b-a}\int_a^b f(x)dx.$

This theorem allows us to find for positive functions a rectangle that has base $[a,b]$ and height $f(c)$ such that the area of the rectangle is the same as the area given by $\int_a^b f(x)dx$. In other words, $f(c)$ is the average function value over $[a,b]$.

**Review Questions**

In problems #1–8, use antiderivatives to compute the definite integral.

1. $\int_4^9 \left(\frac{3}{\sqrt{x}}\right)dx$

2. $\int_0^1 (t - t^2)dt$
3. $\int_2^5 (\frac{1}{\sqrt{x}} + \frac{1}{\sqrt{2}})dx$
4. $\int_0^1 4(x^2 - 1)(x^2 + 1)dx$
5. $\int_2^8 (\frac{4}{x} + x^2 + x)dx$
6. $\int_2^4 (e^{3x})dx$
7. $\int_1^4 \frac{2}{x+3}dx$
8. Find the average value of $f(x) = \sqrt{x}$ over [1, 9].
9. If f is continuous and $\int_1^4 f(x)dx = 9$, show that f takes on the value 3 at least once on the interval [1, 4].
10. Your friend states that there is no area under the curve of $f(x) = \sin x$ on $[0, 2\pi]$ since he computed $\int_0^{2\pi} \sin x\, dx = 0$. Is he correct? Explain your answer.

**Review Answers**

1. $\int_4^9 (\frac{3}{\sqrt{x}})dx = 6$
2. $\frac{1}{6}$
3. $\int_2^5 (\frac{1}{\sqrt{x}} + \frac{1}{\sqrt{2}})dx = 2\sqrt{5} - 2\sqrt{2} + \frac{3\sqrt{2}}{2}$
4. $\int_0^1 4(x^2 - 1)(x^2 + 1)dx = -\frac{16}{5}$
5. $\int_2^8 (\frac{4}{x} + x^2 + x)dx = \frac{9417}{48} \approx 196.19$
6. $\int_2^4 (e^{3x})dx = \frac{e^{12} - e^6}{3}$
7. $\int_1^4 \frac{2}{x+3}dx = 2\ln 7 - 2\ln 4$
8. $\frac{13}{6}$
9. Apply the Mean Value Theorem for integrals.
10. He is partially correct. The definite integral $\int_0^{2\pi} \sin x\, dx$ computes the **net** area under the curve. However, the area between the curve and the x-axis is given by:

$$A = 2\int_0^\pi \sin x\, dx = -\cos x\big]_0^\pi = 2.$$

## 7.6 The Fundamental Theorem of Calculus

**Learning Objectives**
- Use the Fundamental Theorem of Calculus to evaluate definite integrals

**Introduction**

In the Lesson on Evaluating Definite Integrals, we evaluated definite integrals using antiderivatives. This process was much more efficient than using the limit definition. In this lesson we will state the Fundamental Theorem of Calculus and continue to work on methods for computing definite integrals.

***Fundamental Theorem of Calculus:***

Let $f$ be continuous on the closed interval $[a, b]$.

1. If function $F$ is defined by $F(x) = \int_a^x f(t)dx$ on $[a, b]$, then $F'(x) = f(x)$ on $[a, b]$.
2. If $g$ is any antiderivative of $f$ on $[a, b]$, then

$$\int_a^b f(t)dt = g(b) - g(a).$$

We first note that we have already proven part 2 as Theorem 4.1. The proof of part 1 appears at the end of this lesson.

**Think about this Theorem.** Two of the major unsolved problems in science and mathematics turned out to be solved by calculus which was invented in the seventeenth century. These are the ancient problems:

1. Find the areas defined by curves, such as circles or parabolas.
2. Determine an instantaneous rate of change or the slope of a curve at a point.

With the discovery of calculus, science and mathematics took huge leaps, and we can trace the advances of the space age directly to this Theorem.

Let's continue to develop our strategies for computing definite integrals. We will illustrate how to solve the problem of finding the area bounded by two or more curves.

**Example 1:**

Find the area between the curves of $f(x) = x$ and $g(x) = x^3$.

**Solution:**

We first observe that there are no limits of integration explicitly stated here. Hence we need to find the limits by analyzing the graph of the functions.

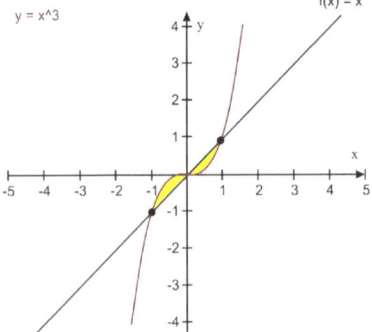

We observe that the regions of interest are in the first and third quadrants from $x = -1$ to $x = 1$. We also observe the symmetry of the graphs about the origin. From this we see that the total area enclosed is

$$2\int_0^1 (x - x^3)dx = 2\left[\int_0^1 x\,dx - \int_0^1 x^3\,dx\right] = 2\left[\frac{x^2}{2}\Big|_0^1 - \frac{x^4}{4}\Big|_0^1\right] = 2\left[\frac{1}{2} - \frac{1}{4}\right] = 2\left[\frac{1}{4}\right] = \frac{1}{2}$$

**Example 2:**

Find the area between the curves of $f(x) = |x - 1|$ and the $x$-axis from $x = -1$ to $x = 3$.

**Solution:**

We observe from the graph that we will have to divide the interval $[-1, 3]$ into subintervals $[-1, 1]$ and $[1, 3]$.

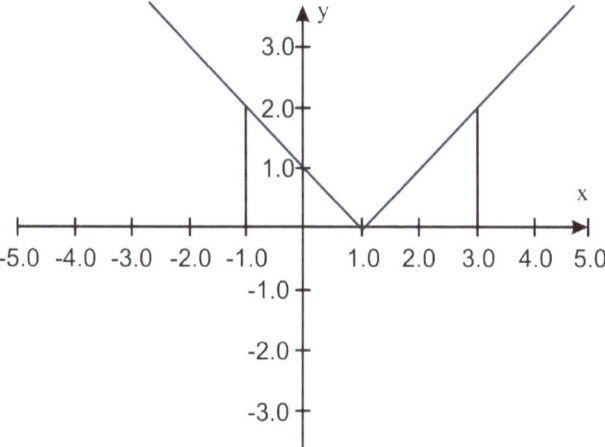

Hence the area is given by

$$\int_{-1}^{1}(-x+1)dx + \int_{1}^{3}(x-1)dx = \left(-\frac{x^2}{2}+x\right)\Big|_{-1}^{+1} + \left(\frac{x^2}{2}-x\right)\Big|_{+1}^{+3} = 2+2 = 4.$$

**Example 3:**
Find the area enclosed by the curves of $f(x) = x^2 + 2x + 1$ and $g(x) = -x^2 - 2x + 1$.

**Solution:**
The graph indicates the area we need to focus on.

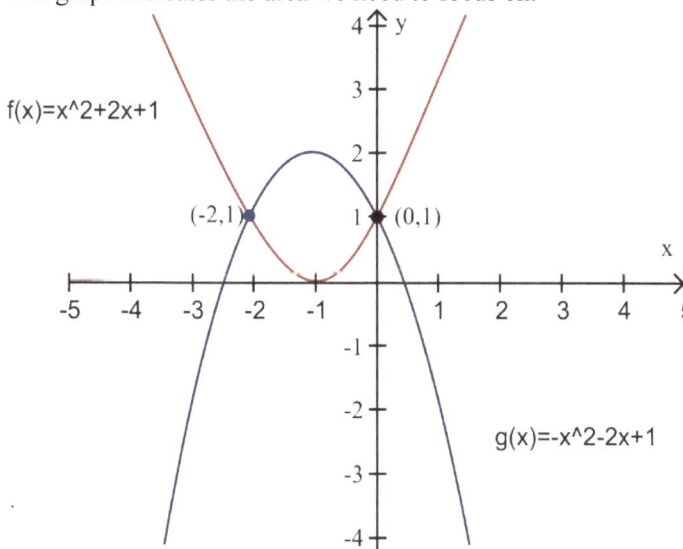

$$\int_{-2}^{0}(-x^2 - 2x + 1)dx - \int_{-2}^{0}(x^2 + 2x + 1)dx = \left(-\frac{x^3}{3} - x^2 + x\right)\Big|_{-2}^{0} + \left(\frac{x^2}{3} - x^3 + x\right)\Big|_{-2}^{0} = \frac{10}{3} - \frac{2}{3} = \frac{8}{3}$$

Before providing another example, let's look back at the first part of the Fundamental Theorem. If function $F$ is defined by $F(x) = \int_{a}^{x} f(t)dt$ on $[a, b]$ then $F'(x) = f(x)$ on $[a, b]$. Observe that if we differentiate the integral with respect to $x$ we have

$$\frac{d}{dx}\int_{a}^{x} f(t)dt = F'(x) = f(x).$$

This fact enables us to compute derivatives of integrals as in the following example.

**Example 4:**
Use the Fundamental Theorem to find the derivative of the following function:
$$g(x) = \int_0^x (1 + \sqrt[3]{t})dt.$$

**Solution:**
While we could easily integrate the right side and then differentiate, the Fundamental Theorem enables us to find the answer very routinely.
$$g'(x) = \frac{d}{dx}\int_0^x (1 + \sqrt[3]{t})dt = 1 + \sqrt[3]{x}.$$

This application of the Fundamental Theorem becomes more important as we encounter functions that may be more difficult to integrate such as the following example.

**Example 5:**
Use the Fundamental Theorem to find the derivative of the following function:
$$g(x) = \int_2^x (t^2 \cos t)dt.$$

**Solution:**
In this example, the integral is more difficult to evaluate. The Fundamental Theorem enables us to find the answer routinely.
$$g'(x) = \frac{d}{dx}\int_2^x (t^2 \cos t)dt = x^2 \cos x.$$

**Lesson Summary**

1. We used the Fundamental Theorem of Calculus to evaluate definite integrals.

***Fundamental Theorem of Calculus***

Let $f$ be continuous on the closed interval $[a, b]$.

1. If function $F$ is defined by $F(x) = \int_a^x f(t)dt$, on $[a, b]$, then $F'(x) = f(x)$ on $[a, b]$.
2. If $g$ is any antiderivative of $f$ on $[a, b]$, then
$$\int_a^b f(t)dt = g(b) - g(a).$$

We first note that we have already proven part 2 as Theorem 4.1.

**Proof of Part 1.**

1. Consider $F(x) = \int_a^x f(t)dt$ on $[a, b]$.
2. $x, c \in [a, b]$, $c < x$.
Then $\int_a^x f(t)dt = \int_a^c f(t)dt + \int_c^x f(t)dt$ by our rules for definite integrals.
3. Then $\int_a^x f(t)dt - \int_a^c f(t)dt = \int_c^x f(t)dt$. Hence $F(x) - F(c) = \int_c^x f(t)dt$.
4. Since $f$ is continuous on $[a, b]$ and $x, c \in [a, b]$, $c < x$ then we can select $u, v \in [c, v]$ such that $f(u)$ is the minimum value of and $f(v)$ is the maximum value of $f$ in $[c, x]$. Then we can consider $f(u)(x - c)$ as a lower sum and $f(v)(x - c)$ as an upper sum of $f$ from $c$ to $x$. Hence
5. $f(u)(x - c) \leq \int_c^x f(t)dt \leq f(v)(x - c)$.
6. By substitution, we have:
$f(u)(x - c) \leq F(x) - F(c) \leq f(v)(x - c)$.
7. By division, we have
$$f(u) \leq \frac{F(x) - F(c)}{x - c} \leq f(v).$$
8. When $x$ is close to $c$, then both $f(u)$ and $f(v)$ are close to $f(c)$ by the continuity of $f$

9. Hence $\lim_{x \to c^+} \frac{F(x)-F(c)}{x-c} = f(c)$. Similarly, if $x < c$ then $\lim_{x \to c^-} \frac{F(x)-F(c)}{x-c} = f(c)$. Hence, $\lim_{x \to c} \frac{F(x)-F(c)}{x-c} = f(c)$.

10. By the definition of the derivative, we have that
$F'(c) = \lim_{x \to c} \frac{F(x)-F(c)}{x-c} = f(c)$ for every $c \in [a, b]$. Thus, $F$ is an antiderivative of $f$ on $[a, b]$.

**Review Questions**

In problems #1–4, sketch the graph of the function f(x) in the interval [a, b]. Then use the Fundamental Theorem of Calculus to find the area of the region bounded by the graph and the x-axis. (Hint: Examine the graph of the function and divide the interval accordingly.)

1. $f(x) = 2x + 3, [0, 4]$
2. $f(x) = e^x, [0, 2]$
3. $f(x) = x^2 + x, [1, 3]$
4. $f(x) = x^2 - x, [0, 2]$

In problems #5–7 use antiderivatives to compute the definite integral. (Hint: Examine the graph of the function and divide the interval accordingly.)

5. $\int_{-1}^{+1} |x| \, dx$
6. $\int_0^3 |x^3 - 1| \, dx$
7. $\int_{-2}^{+4} [|x-1| + |x+1|] \, dx$

In problems #8–10, find the area between the graphs of the functions.

8. $f(x) = \sqrt{x}, g(x) = x$
9. $f(x) = x^2, g(x) = 4$
10. $f(x) = x^2 + 1, g(x) = 3 - x$, on the interval (0, 3) (Hint: you will need to add 2 integrals)

**Review Answers**

1. Area $= 28$

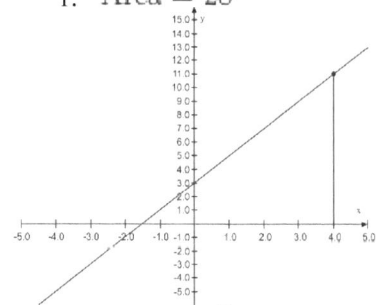

2. Area $= e^2 - 1$

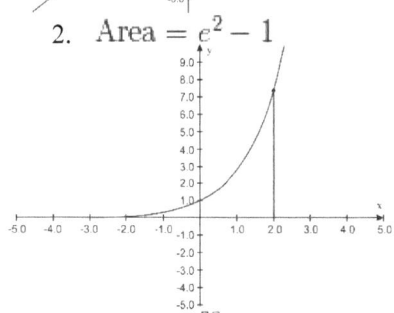

3. Area $= \frac{38}{3}$

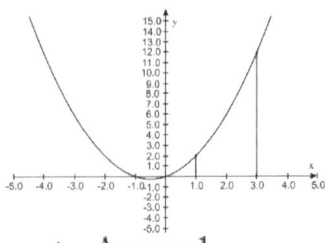

4. Area = 1

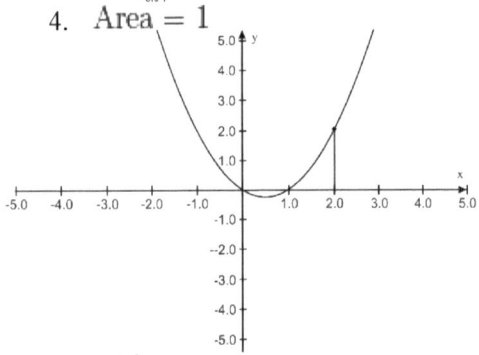

5. $\int_{-1}^{+1} |x|\,dx = 1$
6. 18.75
7. $\int_{-2}^{+4} \left[|x-1| + |x+1|\right] dx = 22$
8. Area = 1/6

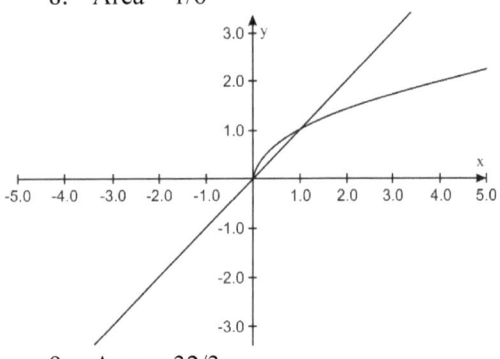

9. Area = 32/3

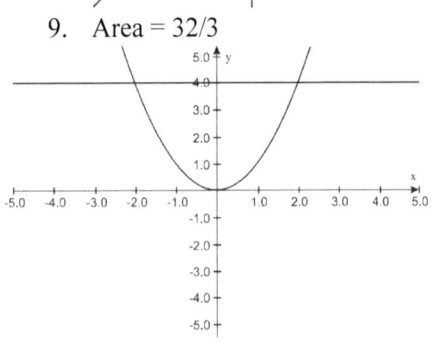

10. Area = $\dfrac{59}{6}$

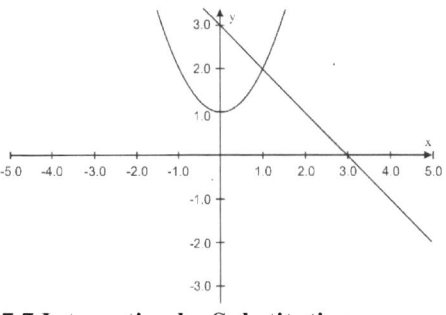

## 7.7 Integration by Substitution
**Learning Objectives**
- Integrate composite functions
- Use change of variables to evaluate definite integrals
- Use substitution to compute definite integrals

**Introduction**

In this lesson we will expand our methods for evaluating definite integrals. We first look at a couple of situations where finding antiderivatives requires special methods. These involve finding antiderivatives of composite functions and finding antiderivatives of products of functions.

*Antiderivatives of Composites*

Suppose we needed to compute the following integral:

$$\int 3x^2 \sqrt{1+x^3}\, dx.$$

Our rules of integration are of no help here. We note that the integrand is of the form $f(g(x)) * g'(x)$ where $g(x) = 1 + x^3$ and $f(x) = \sqrt{x}$.

Since we are looking for an antiderivative $F$ of $f$, and we know that $F' = f$, we can re-write our integral as

$$\int \sqrt{1+x^3} \cdot 3x^2\, dx = \frac{2}{3}(\sqrt{1+x^3})^{\frac{3}{2}} + C.$$

In practice, we use the following substitution scheme to verify that we can integrate in this way:

4. Integrate with respect to $u$:

$\int \sqrt{1+x^3} \cdot 3x^2\, dx = \int \sqrt{u}\, du$, where $u = 1 + x^3$ and $du = 3x^2\, dx$.

3. Change the original integral in $x$ to an integral in $u$:

2. Differentiate both sides so $du = 3x^2\, dx$.

1. Let $u = 1 + x^3$.

$$\int \sqrt{u}\, du = \int u^{\frac{1}{2}}\, du = \frac{2}{3}u^{\frac{3}{2}} + C.$$

5. Change the answer back to $x$:

While this method of substitution is a very powerful method for solving a variety of problems, we will find that we sometimes will need to modify the method slightly to address problems, as in the following example.

**Example 1:**

Compute the following indefinite integral:

$$\int x^2 e^{x^3}\, dx.$$

**Solution:**

We note that the derivative of $x^3$ is $3x^2$; hence, the current problem is not of the form $\int F'(g(x)) \cdot g'(x)\, dx$. But we notice that the derivative is off only by a constant of $3$ and we know that constants are easy to deal with when differentiating and integrating. Hence

Let $u = x^3$.

Then $du = 3x^2 dx$.
Then $\frac{1}{3}du = x^2 dx$, and we are ready to change the original integral from $x$ to an integral in $u$ and integrate:

$$\int x^2 e^{x^3} dx = \int e^u \left(\frac{1}{3}du\right) = \frac{1}{3}\int e^u du = \frac{1}{3}e^u + C.$$

Changing back to $x$, we have

$$\int x^2 e^{x^3} dx = \frac{1}{3}e^{x^3} + C.$$

We can also use this substitution method to evaluate definite integrals. If we attach limits of integration to our first example, we could have a problem such as

$$\int_1^4 \sqrt{1+x^3} \cdot 3x^2 dx.$$

The method still works. However, we have a choice to make once we are ready to use the Fundamental Theorem to evaluate the integral.

Recall that we found that $\int \sqrt{1+x^3} \cdot 3x^2 dx = \int \sqrt{u}\, du$ for the indefinite integral. At this point, we could evaluate the integral by changing the answer back to $x$ or we could evaluate the integral in $u$. But we need to be careful. Since the original limits of integration were in $x$, we need to change the limits of integration for the equivalent integral in $u$. Hence,

$\int_1^4 \sqrt{1+x^3} \cdot 3x^2 dx = \int_{u=2}^{65} \sqrt{u}\, du$, where $u = 1 + x^3$

$$\int_1^4 \sqrt{1+x^3} \cdot 3x^2 dx = \int_{u=2}^{65} \sqrt{u}\, du = \frac{2}{3}u^{\frac{3}{2}}\Big|_{u=2}^{u=65} = \frac{2}{3}(\sqrt{65^3} - \sqrt{8}).$$

### Integrating Products of Functions

We are not able to state a rule for integrating products of functions, $\int f(x)g(x)dx$ but we can get a relationship that is almost as effective. Recall how we differentiated a product of functions:

$$\frac{d}{dx}f(x)g(x) = f(x)g'(x) + g(x)f'(x).$$

So by integrating both sides we get
$\int [f(x)g'(x) + g(x)f'(x)]dx = f(x)g(x)$, or

$$\int f(x)g'(x)dx = f(x)g(x) - \int g(x)f'(x).$$

In order to remember the formula, we usually write it as

$$\int u\, dv = uv - \int v\, du.$$

We refer to this method as integration by parts. The following example illustrates its use.

**Example 2:**
Use integration by parts method to compute

$$\int xe^x dx.$$

**Solution:**
We note that our other substitution method is not applicable here. But our integration by parts method will enable us to reduce the integral down to one that we can easily evaluate.
Let $u = x$ and $dv = e^x dx$ then $du = dx$ and $v = e^x$
By substitution, we have

$$\int xe^x dx = xe^x - \int e^x dx.$$

We can easily evaluate the integral and have

$$\int xe^x dx = xe^x - \int e^x dx = xe^x - e^x + C.$$

And should we wish to evaluate definite integrals, we need only to apply the Fundamental Theorem to the antiderivative.

**Lesson Summary**
1. We integrated composite functions.
2. We used change of variables to evaluate definite integrals.
3. We used substitution to compute definite integrals.

**Review Questions**

Compute the integrals in problems #1–8.
1. $\int x \ln x \, dx$
2. $\int \frac{x}{\sqrt{2x+1}} dx$
3. $\int_0^1 x^3 \sqrt{1-x^2} \, dx$
4. $\int x \cos x \, dx$
5. $\int_0^1 x^2 \sqrt{x^3+9} \, dx$
6. $\int \left(\frac{1}{x^2} \cdot e^{\frac{1}{x}}\right) dx$
7. $\int x^3 e^{x^2} dx$
8. $\int_1^e \frac{1}{x} dx$

**Review Answers**

1. $\int x \ln x \, dx = \frac{x^2(2\ln x - 1)}{4} + C$
2. $\int \frac{x}{\sqrt{2x+1}} dx = \frac{(x-1)\sqrt{2x+1}}{3} + c$
3. $\int_0^1 x^3 \sqrt{1-x^2} \, dx = \frac{2}{15}$
4. $\int x \cos x \, dx = x \sin x + \cos x + c$
5. $\int_0^1 x^2 \sqrt{x^3+9} \, dx = \frac{2}{9}\left[10^{\frac{3}{2}} - 27\right]$
6. $\int \left(\frac{1}{x^2} \cdot e^{\frac{1}{x}}\right) dx = -e^{\frac{1}{x}} + c$
7. $\int x^3 e^{x^2} dx = \frac{1}{2}e^{x^2}(x^2-1) + c$
8. $\int_1^e \frac{1}{x} dx = 1$

**Integration by Substitution Practice**

Find the indefinite integral.

1. $\int (x^2-1)^3 (2x)\,dx$
2. $\int \sqrt{3-x^3}\,(3x^2)\,dx$
3. $\int (x-3)^{5/2}\,dx$
4. $\int x(1-2x^2)^3\,dx$
5. $\int \dfrac{x^2}{(x^3-1)^2}\,dx$
6. $\int \dfrac{6x}{(1+x^2)^3}\,dx$
7. $\int \dfrac{4x+6}{(x^2+3x+7)^3}\,dx$
8. $\int m^3\sqrt{m^4+2}\,dm$
9. $\int \dfrac{x^2}{\sqrt{1-x^3}}\,dx$
10. $\int \dfrac{t+2t^2}{\sqrt{t}}\,dt$
11. $\int \left(1+\dfrac{1}{t}\right)^3 \left(\dfrac{1}{t^2}\right)dt$
12. $\int \dfrac{1}{3x^2}\,dx$
13. $\int (3-2x-4x^2)(1+4x)\,dx$

14. Find the equation of the function $f$ whose graph passes through the point $\left(0,\dfrac{7}{3}\right)$ and whose derivative is $f'(x) = x\sqrt{1-x^2}$.

Answers:

1. $\dfrac{(x^2-1)^4}{4}+C$
2. $\dfrac{-2}{3}(3-x^3)^{3/2}+C$
3. $\dfrac{2}{7}(x-3)^{7/2}+C$
4. $\dfrac{-1}{16}(1-2x^2)^4+C$
5. $\dfrac{-1}{3(x^3-1)}+C$
6. $-\dfrac{3}{2(1+x^2)^2}+C$
7. $\dfrac{-1}{(x^2+3x+7)^2}+C$
8. $\dfrac{1}{6}(m^4+2)^{3/2}+C$
9. $\dfrac{-2}{3}\sqrt{1-x^3}+C$
10. $\dfrac{2}{15}t^{3/2}(5+6t)+C$
11. $\dfrac{-1}{4}\left(1+\dfrac{1}{t}\right)^4+C$
12. $\dfrac{-1}{3x}+C$
13. $\dfrac{-1}{4}(3-2x-4x^2)^2+C$
14. $f(x)=\dfrac{-1}{3}(1-x^2)^{3/2}+\dfrac{8}{3}$

## More Substitution Practice

Integrate!

1.) $\int 2x(x^2-1)^3 \, dx$

2.) $\int 3x^2 \sqrt{3-x^3} \, dx$

3.) $\int (x-3)^{\frac{5}{2}} \, dx$

4.) $\int x(1-2x^2)^3 \, dx$

5.) $\int \dfrac{x^2}{(x^3-1)^2} \, dx$

6.) $\int \dfrac{4x+6}{(x^2+3x+7)^3} \, dx$

7.) Find the equation of the function $f(x)$ whose graph passes through the point $\left(0, \dfrac{7}{3}\right)$ and whose derivative is $f'(x) = x\sqrt{1-x^2}$.

Answers:

1.) $\dfrac{(x^2-1)^4}{4} + C$

2.) $-\dfrac{2}{3}(3-x^3)^{\frac{3}{2}} + C$

5.) $-\dfrac{1}{3(x^3-1)} + C$

6.) $-\dfrac{1}{(x^2+3x+7)^2} + C$

3.) $\dfrac{2}{7}(x-3)^{\frac{7}{2}} + C$

4.) $-\dfrac{1}{16}(1-2x^2)^4 + C$

7.) $f(x) = -\dfrac{1}{3}(1-x^2)^{\frac{3}{2}} + \dfrac{8}{3}$

## More Substitution Practice

Integrate! Some of these require the general power rule (substitution), others do not.

1.) $\displaystyle\int \frac{6x}{(1+x^2)^3}\,dx$

2.) $\displaystyle\int \frac{t+2t^2}{\sqrt{t}}\,dt$

3.) $\displaystyle\int (3-2x-4x^2)(1+4x)\,dx$

4.) $\displaystyle\int \frac{1}{3x^2}\,dx$

5.) $\displaystyle\int 5x\sqrt[3]{1-x^2}\,dx$

6.) $\displaystyle\int (x+5)^6\,dx$

7.) $\displaystyle\int (3x-8)^{10}\,dx$

Answers:

1.) $-\dfrac{3}{2(1+x^2)^2}+C$

2.) $\dfrac{2}{15}t^{\frac{3}{2}}(5+6t)+C$

3.) $-\dfrac{1}{4}(3-2x-4x^2)^2+C$

5.) $-\dfrac{15}{8}(1-x^2)^{\frac{4}{3}}+C$

6.) $\dfrac{(x+5)^7}{7}+C$

7.) $\dfrac{(3x-8)^{11}}{33}+C$

4.) $-\dfrac{1}{3x}+C$

## 7.8 Numerical Integration

**Learning Objectives**
- Use the Trapezoidal Rule to solve problems
- Estimate errors for the Trapezoidal Rule
- Use Simpson's Rule to solve problems
- Estimate Errors for Simpson's Rule

**Introduction**

Recall that we used different ways to approximate the value of integrals. These included Riemann Sums using left and right endpoints, as well as midpoints for finding the length of each rectangular tile. In this lesson we will learn two other methods for approximating integrals. The first of these, the Trapezoidal Rule, uses areas of trapezoidal tiles to approximate the integral. The second method, Simpson's Rule, uses parabolas to make the approximation.

***Trapezoidal Rule***

Let's recall how we would use the midpoint rule with $n=4$ rectangles to approximate the area under the graph of $f(x) = x^2 + 1$ from $x=0$ to $x=1$.

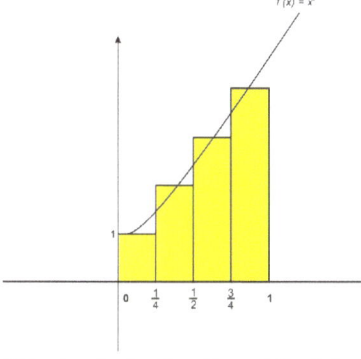

If instead of using the midpoint value within each sub-interval to find the length of the corresponding rectangle, we could have instead formed trapezoids by joining the maximum and minimum values of the function within each sub-interval:

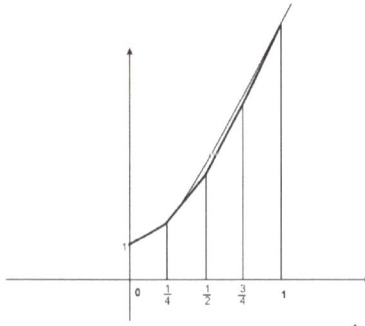

The area of a trapezoid is $A = \frac{h(b_1+b_2)}{2}$, where $b_1$ and $b_2$ are the lengths of the parallel sides and $h$ is the height. In our trapezoids the height is $\triangle x$ and $b_1$ and $b_2$ are the values of the function. Therefore in finding the areas of the trapezoids we actually average the left and right endpoints of each sub-interval. Therefore a typical trapezoid would have the area

$$A = \frac{\triangle x}{2}\left(f(x_{i-1}) + f(x_i)\right).$$

To approximate $\int_a^b f(x)dx$ with $n$ of these trapezoids, we have

$$\int_a^b f(x)dx \approx \frac{1}{2}\left[\sum_{i=1}^n f(x_{i-1})\Delta x + \sum_{i=1}^n f(x_i)\Delta x\right]$$

$$= \frac{\Delta x}{2}[f(x_0) + f(x_1) + f(x_1) + f(x_2) + f(x_2) + \ldots + f(x_{n-1})f(x_n)]$$

$$= \frac{\Delta x}{2}[f(x_0) + 2f(x_1) + 2f(x_2) + \ldots + 2f(x_{n-1})f(x_n)], \Delta x = \frac{b-a}{n}.$$

**Example 1:**

Use the Trapezoidal Rule to approximate $\int_0^3 x^2 dx$ with $n = 6$.

**Solution:**

We find $\Delta x = \frac{b-a}{n} = \frac{3-0}{6} = \frac{1}{2}$.

$$\int_0^3 x^2 dx \approx \frac{1}{4}\left[f(0) + 2f(\tfrac{1}{2}) + 2f(1) + 2f(\tfrac{3}{2}) + 2f(2) + 2f(\tfrac{5}{2}) + f(3)\right]$$

$$= \frac{1}{4}\left[0 + (2 \cdot \tfrac{1}{4}) + (2 \cdot 1) + (2 \cdot \tfrac{9}{4}) + (2 \cdot 4) + (2 \cdot \tfrac{25}{4}) + 9\right]$$

$$= \frac{1}{4}\left[\tfrac{73}{2}\right] = \frac{73}{8} = 9.125.$$

Of course, this estimate is not nearly as accurate as we would like. For functions such as $f(x) = x^2$, we can easily find an antiderivative with which we can apply the Fundamental Theorem that

$$\int_0^3 x^2 dx = \left.\frac{x^3}{3}\right|_0^3 = 9.$$

But it is not always easy to find an antiderivative. Indeed, for many integrals it is impossible to find an antiderivative. Another issue concerns the questions about the accuracy of the approximation. In particular, how large should we take n so that the Trapezoidal Estimate for $\int_0^3 x^2 dx$ is accurate to within a given value, say $0.001$? As with our Linear Approximations in the Lesson on Approximation Errors, we can state a method that ensures our approximation to be within a specified value.

***Error Estimates for Simpson's Rule***

We would like to have confidence in the approximations we make. Hence we can choose $n$ to ensure that the errors are within acceptable boundaries. The following method illustrates how we can choose a sufficiently large $n$.

Suppose $|f''(x)| \leq k$ for $a \leq x \leq b$. Then the error estimate is given by

$$|Error_{Trapezoidal}| \leq \frac{k(b-a)^3}{12n^2}.$$

**Example 2:**

Find $n$ so that the Trapezoidal Estimate for $\int_0^3 x^2 dx$ is accurate to $0.001$.

**Solution:**

We need to find $n$ such that $|Error_{Trapezoidal}| \leq 0.001$. We start by noting that $|f''(x)| = 2$ for $0 \leq x \leq 3$. Hence we can take $K = 2$ to find our error bound.

$$|Error_{Trapezoidal}| \leq \frac{2(3-0)^3}{12n^2} = \frac{54}{12n^2}.$$

We need to solve the following inequality for $n$:

$$\frac{54}{12n^2} < 0.001,$$
$$n^2 > \frac{54}{12(0.001)},$$
$$n > \sqrt{\frac{54}{12(0.001)}} \approx 67.08.$$

Hence we must take $n = 68$ to achieve the desired accuracy.

From the last example, we see one of the weaknesses of the Trapezoidal Rule—it is not very accurate for functions where straight line segments (and trapezoid tiles) do not lead to a good estimate of area. It is reasonable to think that other methods of approximating curves might be more applicable for some functions. ***Simpson's Rule*** is a method that uses parabolas to approximate the curve.

***Simpson's Rule***:

As was true with the Trapezoidal Rule, we divide the interval $[a, b]$ into $n$ sub-intervals of length $\Delta x = \frac{b-a}{n}$. We then construct parabolas through each group of three consecutive points on the graph. The graph below shows this process for the first three such parabolas for the case of $n = 6$ sub-intervals. You can see that every interval except the first and last contains two estimates, one too high and one too low, so the resulting estimate will be more accurate.

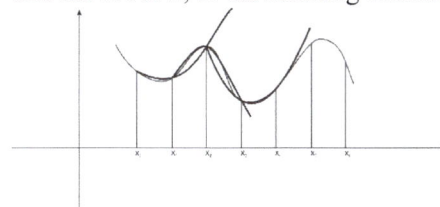

Using parabolas in this way produces the following estimate of the area from Simpson's Rule:

$$\int_a^b f(x)dx \approx \frac{\Delta x}{3}[f(x_0) + 4f(x_1) + 2f(x_2) + 4f(x_3) + 2f(x_4) \ldots + 2f(x_{n-2}) + 4f(x_{n-1}) + f(x_n)].$$

We note that it has a similar appearance to the Trapezoidal Rule. However, there is one distinction we need to note. The process of using three consecutive $x_i$ to approximate parabolas will require that we assume that $n$ must always be an even number.

***Error Estimates for the Trapezoidal Rule***

As with the Trapezoidal Rule, we have a formula that suggests how we can choose $n$ to ensure that the errors are within acceptable boundaries. The following method illustrates how we can choose a sufficiently large $n$.

Suppose $|f^4(x)| \leq k$ for $a \leq x \leq b$. Then the error estimate is given by

$$|Error_{simpson}| \leq \frac{k(b-a)^5}{180n^4}.$$

**Example 3:**

a. Use Simpson's Rule to approximate $\int_1^4 \frac{1}{x}dx$ with $n = 6$.

**Solution:**

We find $\Delta x = \frac{b-a}{n} = \frac{4-1}{6} = \frac{1}{2}$.

$$\int_1^4 \frac{1}{x} dx \approx \frac{1}{6}\left[f(1) + 4f(\tfrac{3}{2}) + 2f(2) + 4f(\tfrac{5}{2}) + 2f(3) + 4f(\tfrac{7}{2}) + f(4)\right]$$

$$= \frac{1}{6}\left[1 + (4 \cdot \tfrac{2}{3}) + (2 \cdot \tfrac{1}{2}) + (4 \cdot \tfrac{2}{5}) + (2 \cdot \tfrac{1}{3}) + (4 \cdot \tfrac{2}{7}) + \tfrac{1}{4}\right]$$

$$= \frac{1}{6}\left[\tfrac{3517}{420}\right] = 1.3956.$$

This turns out to be a pretty good estimate, since we know that

$$\int_1^4 \frac{1}{x} dx = \ln x \Big]_1^4 = \ln(4) - \ln(1) = 1.3863.$$

Therefore the error is less than $0.01$.

b. Find $n$ so that the Simpson Rule Estimate for $\int_1^4 \frac{1}{x}dx$ is accurate to $0.001$.

**Solution:**

We need to find $n$ such that $|Error_{simpson}| \leq 0.001$. We start by noting that $|f^4(x)| = \left|\frac{24}{x^5}\right|$ for $1 \leq x \leq 4$. Hence we can take $K = 24$ to find our error bound:

$$|Error_{simpson}| \leq \frac{24(4-1)^5}{180n^4} = \frac{5832}{180n^4}.$$

Hence we need to solve the following inequality for $n$:

$$\frac{5832}{180n^4} < 0.001.$$

We find that

$$n^4 > \frac{5832}{180(0.001)},$$

$$n > \sqrt[4]{\frac{5832}{180(0.001)}} \approx 13.42.$$

Hence we must take $n = 14$ to achieve the desired accuracy.

**Technology Note: Estimating a Definite Integral with a TI-83/84 Calculator**

We will estimate the value of $\int_1^4 \frac{1}{x}dx$.

1. Graph the function $f(x) = \frac{1}{x}$ with the [WINDOW] setting shown below.
2. The graph is shown in the second screen.
3. Press **2nd [CALC]** and choose option **7** (see menu below)
4. When the fourth screen appears, press **[1] [ENTER]** then **[4] [ENTER]** to enter the lower and upper limits.
5. The final screen gives the estimate, which is accurate to 7 decimal places.

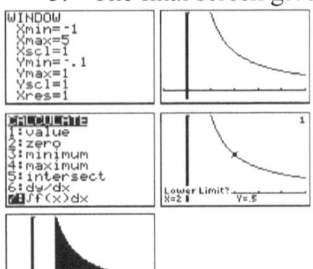

**Lesson Summary**

1. We used the Trapezoidal Rule to solve problems.
2. We estimated errors for the Trapezoidal Rule.
3. We used Simpson's Rule to solve problems.

4. We estimated Errors for Simpson's Rule.

**Important notes regarding this book**

**Dear learner**

**At the end of the book of study you should know**

- ✓ **The basics of calculus and its theories.**
- ✓ **Some arguments posed by this field.**
- ✓ **Possibilities of accuracy and proximities.**
- ✓ **Philosophies of mathematics and its infinity**

With such skills and knowledge . we then say welcome to the controversies of mathematics and science.

**Thank you.**

# Dedication

To my mother Ms Dorah Sello and father Mr. NR Sello

                                              Thank you

                                                    Mr. T.V Sello

www.ingramcontent.com/pod-product-compliance
Lightning Source LLC
Chambersburg PA
CBHW051147220526
45473CB00003B/693